T0195573

Original

ZEN

Dwelling

BUDDHIST

Place

ESSAYS

Also by Robert Aitken

The Dragon Who Never Sleeps

Encouraging Words

The Gateless Barrier

The Mind of Clover

The Practice of Perfection

Taking the Path of Zen

A Zen Wave

(with David Steindl-Rast)
The Ground We Share

Original

Z E N

Dwelling

B U D D H I S T

Place

E S S A Y S

Robert Aitken

COUNTERPOINT

First paperback edition 1997

Many of these essays have appeared, in slightly different form, in the following
journals: *Blind Donkey*; *Buddhist-Christian Studies*; *The Eastern Buddhist:
New Series*; *Mind Moon Circle*; *Tricycle: The Buddhist Review*; *Turning Wheel*;
The Wallace Stevens Journal.

Library of Congress Cataloging-in-Publication Data
Aitken, Robert, 1917–
Original dwelling place: Zen Buddhist essays / Robert Aitken.
1. Spiritual life—Zen Buddhism.   2. Zen Buddhism—Doctrines.
I. Title.
BQ9288.A356   1996
294.3'927—dc20   95-51453
ISBN 1-887178-41-4 (alk. paper)

ISBN: 978-1-887178-41-9

Printed in the United States of America

Designer: David Bullen
Compositor: Wilsted & Taylor

COUNTERPOINT
2560 Ninth Street, Suite 318
Berkeley, CA 94710
www.counterpointpress.com

Contents

Acknowledgments

IN earlier drafts, many of these talks, essays, and prefaces were machine-copied, bound, and distributed privately to libraries of Diamond Sangha centers, and to prison libraries in the State of Hawai'i. David Steinkraus helped me to choose the pieces in that collection and to arrange them. His work established the pattern of the present volume. Jason Binford, Olyn Garfield, and Shakti Murthy helped me to develop the collection. Ms. Murthy did subsequent selecting and arranging, and provided me with especially useful editorial assistance.

"The Way of Dōgen Zenji" is from *Dōgen Kigen: Mystical Realist* by Hee-Jin Kim. Copyright © 1987 by University of Arizona Press. Reprinted by permission.

"Nyogen Senzaki: An American Hotei" is from *Buddhism and Zen* by Nyogen Senzaki and Ruth Strout McCandless. Copyright © 1987 by North Point Press. Reprinted by permission of Farrar Straus & Giroux.

The poem "People's Abuse" by Musō Soseki is from *Sun at Midnight: Poems and Sermons by Musō Soseki*, translated by W. S. Merwin and Sōiku Shigematsu. Copyright © 1989 by North Point Press. Reprinted by permission of Farrar Straus & Giroux.

The excerpt from the poem "Aubade" is from *Philip Larkin: Collected Poems*, edited by Anthony Thwaite. Copyright © 1988 by Farrar Straus & Giroux and The Marvell Press. Reprinted by permission of Farrar Straus & Giroux.

Over the years, I presented talks and essays in this collection in Diamond Sangha classes and incorporated ideas emerging from the discussions into the final drafts. I have also used suggestions sent in by correspondents after they read some of these pieces in their earlier published forms.

I am, as always, grateful to Jack Shoemaker, master editor, to Carole McCurdy and the staff of Counterpoint, and also to Nancy Palmer Jones, for their steadfast and talented guidance.

R. A.

Strip off the blinders, unload the saddlebags!

HSÜEH-TOU CH'UNG-HSIEN

Unscrew the locks from the doors!
Unscrew the doors themselves from their jambs!

WALT WHITMAN

To the living presence of
R. H. Blyth

Original
Dwelling
Place

ANCESTORS

Introduction

WHEN I first took up the Zen Buddhist way, I noticed that my various teachers would pepper their talks with quotations from their teachers. In subsequent reading, I found that masters of old times would do the same. Now that I am a teacher, I find an intimate rationale for this practice. All of my guides have passed away, but they are alive in my mind and body.

By way of leading off this collection of essays, then, I present memorials of my first two teachers: the monk Nyogen Senzaki, who introduced me to formal Zen Buddhist practice in 1947 and guided me until his death eleven years later, and his friend and colleague Nakagawa Sōen Rōshi, who was my teacher from 1950 until 1957 and an important adviser thereafter. It also includes memorials of two Zen friends and mentors: Dr. R. H. Blyth, scholar of Japanese poetry and humor, with whom I was interned during part of World War II, and who was an important guide until his death in 1964, and Dr. D. T. Suzuki, my instructor at the University of Hawai'i in 1949–50, who was likewise a valued guide until his death in 1966.

These four teachers, together with my two later teachers, Yasutani Haku'un Rōshi and Yamada Kōun Rōshi, formed my character and my teaching, and have, in one way or another, influenced the development of the Diamond Sangha as a stream of North and South American, Australasian, and European Zen Buddhism.

Finally, I have included an essay on the Reverend Dwight Goddard, an American pilgrim in the fields of Asian Buddhism, and Zen in particular, who broke stony ground for the rest of us. He was a pioneer in Buddhist-Christian studies, and his collection of translations, *A Buddhist Bible*, has had a profound influence on the development of Zen Buddhism in the English-speaking world. I feel a special affinity for this Bodhisattva of Thetford, Vermont, who knew the Dharma when he saw it, even in mangled translation and garbled interpretation.

The linguistic and cultural barriers the Reverend Mr. Goddard faced are still firmly in place, though perhaps they have become a little porous. Still, Japan continues to be Japan; the Americas and Europe are very much themselves. I am profoundly grateful to my teachers who guided me to the place where I can, in a milieu very different from theirs, at least begin to apply the essence of their teaching, an essence that has no race or nation.

Nyogen Senzaki

An American Hotei

NYOGEN Senzaki was born on the Siberian peninsula of Kamchatka in 1876 of a Japanese mother and an unknown father. He used to joke that he was probably half Chinese, and indeed he looked rather more Chinese than Japanese. But he himself did not know.

His mother died at his birth, and he was adopted by a Japanese Tendai Buddhist priest who may have been ministering to expatriate Japanese in Siberia. They moved to Japan where the boy grew up and began medical studies, but his education was cut short and his life was completely changed by his adoptive father's sudden death. At this tragic loss, young Senzaki renounced the world and became a Zen Buddhist monk, first in the Sōtō school, then in the Rinzai monastery Enkakuji in Kitakamakura.

There was no mother figure in young Senzaki's life. He told me that as a boy he tried to visualize the mother who bore him but could only summon up a vague outline. His adoptive father, he said, was an important moral and religious influence:

My foster father began to teach me Chinese classics when I was five years old. He was a Kegon [Hua-yen] scholar so he naturally gave me training in Buddhism. When I was eighteen years old, I had finished reading the Chinese Tripitaka, but now in this old age I do not remember what I read. Only his influence remains: to live up to the Buddhist ideals outside of name and fame and to avoid as far as possible the world of loss and gain.[1]

Renouncing the world might have seemed the ultimate fulfillment of his father's teaching, but the young monk found himself in institutional religion, with a hierarchy of titles and authority that was worldly indeed. He loved his teacher, but he came to reject what he called "cathedral Zen." Reminiscing with his American students some thirty years later, he remarked:

> When my master was alive, I asked him to excuse me from all official ranks and titles of our church, and allow me to walk freely in the streets of the world. I do not wish to be called Reverend, Bishop, or by any other church title. To be a member of the great American people and walk any stage of life as I please, is honorable enough for me. I want to be an American Hotei, a happy Jap in the streets.[2]

Hotei is the so-called Laughing Buddha, a legendary figure who wandered about begging for cakes and fruits and then giving them to children. When he met monks, he would challenge them on a point of realization. "He is my ideal Zen teacher," Senzaki said.

> I do not mean his stoutness, nor his life as a street wanderer, nor his deeds as Santa Claus, but his anti-church idea. Churches are all right as long as they carry the true teaching, but when they start commercializing themselves, they spoil the teaching more than anything else.[3]

Senzaki's Zen master was Shaku Sōen,[4] who introduced Zen Buddhism to the United States at the World Parliament of Religions in Chicago in 1893 and who was teacher also of D. T. Suzuki. He gave Senzaki permission to leave the monastery before his formal training was completed, providing him with a remarkable "to whom it may concern" letter of approval, dated autumn 1901:

> Monk Nyogen tries to live the Bhikkhu's life according to the teaching of Buddha, to be nonsectarian with no connection to a temple or headquarters; therefore, he keeps no property

of his own, refuses to hold a position in the priesthood, and conceals himself from noisy fame and glory. He has, however, the Four Vows—greater than worldly ambition, with Dharma treasures higher than any position, and loving-kindness more valuable than temple treasures.[5]

Senzaki's wanderings took him to northern Japan, where he became priest of a little temple and director of its kindergarten. These were perhaps his happiest days. Fifty years after he left Japan for his life in the United States, he was offered a chance to return for a visit. He accepted this invitation largely because he wanted to see the children he had played with in Aomori, some of whom were already grandparents. Those former students who could be reached held a reunion with him, and when he returned to Los Angeles he told us of recognizing them and calling them by name.

His happy days with the children were also days that established his way of teaching. When he set up a Zen center in San Francisco, he called it the "Mentorgarten," explaining:

I coined the word "Mentorgarten" as I thought the whole world was a beautiful garden where all friends could associate peacefully and be mentors of one another. I took the German *garten* instead of "garden" in English, for I was fond of Froebel's theory of kindergarten and thought that we were all children of the Buddha. . . . As in a kindergarten we had no teacher, but we encouraged one another and tried our best to grow up naturally. And like a nurse of the kindergarten, I sometimes presumed myself as a gardener to do all sorts of labor, but I never forgot that I myself was also a flower of the garden, mingling with old and new friends. . . . I was always happy in this Mentorgarten, and why will I not be so in the future? This is the . . . spirit of the Sangha in primitive Buddhism, nay, not only in primitive Buddhism but in modern Buddhism, so far as it is true Buddhism.[6]

The Russo-Japanese War in 1905 interrupted Senzaki's idyll in his garden of children, and he spoke out against it strongly. This was a time of national pride and jingoism, and word of Senzaki's dangerous behavior reached his teacher. Shaku Sōen had been invited to San Francisco to give instruction in Zen practice by friends he had made in Chicago in 1893, so he asked Senzaki to accompany him as his attendant.[7] They stayed several months, and when Shaku Sōen returned to Japan, Senzaki remained in San Francisco.

"Don't try to teach for twenty years," Shaku Sōen advised his student. Senzaki therefore began his American career as a houseboy and a cook, managing his own short-order restaurant for a while. He studied English and Western philosophy diligently, particularly the works of Immanuel Kant. "I like Kant," he once said to me. "All he needed was a good kick in the pants."

Senzaki also taught the Japanese language during this period, and when he saved enough money, he would hire a hall and give a lecture on Buddhism. He took part in various Japanese cultural events as well, and thus gradually established a foundation for his career of teaching Zen.

At last in 1925 he completed his long apprenticeship and began teaching in his "floating zendō," meeting in homes and apartments of members. Later, in 1928, with help from friends in Japan as well as in San Francisco, he rented an apartment on Bush Street and founded his first center, which was also the first Zen center in the United States.

Though he wanted his students to teach themselves and each other as much as possible, he sought very soon to bring a teacher from Japan to help with their guidance. He succeeded in bringing Furukawa Gyōdō Rōshi to San Francisco for a visit, but this was, I gather, a disastrous venture.

Gyōdō Rōshi had been his fellow monk and was now abbot of Enkakuji—and from the beginning of his visit, he was not amused. Senzaki told me of meeting his ship. There was the rōshi at the rail, resplendent in his robes, and there was Senzaki on the dock, dressed in work pants and a shirt open at the neck. Running along the dock and waving his arms, Senzaki shouted up to his old friend happily, "Dō San!" using the abbreviation of Gyōdō Rōshi's name that he had always used when they were monks together. It was not the formal greeting that the distinguished rōshi had learned to expect.

There were other unpleasant surprises as well, we can be sure, and before long, the rōshi was on his way back to Japan. Senzaki then moved to Los Angeles. I have the impression that his move was related to the rōshi's visit, but I don't know the details. In any case, by 1932, Senzaki was living on Turner Street in Los Angeles, alongside the switching yard of the railroad station—the noisiest, dirtiest, and of course the least expensive part of the city.

It was here, in the course of getting acquainted in the Japanese community, that Senzaki met Mrs. Kin Tanahashi, who was to become perhaps his most important friend. Mrs. Tanahashi and her husband had a small business, and she could not afford to hire someone to look after her son Jimmy, who was mentally disabled. Senzaki offered to care for the boy and refused any payment. Years later, he told me how much he had enjoyed playing with the child and what delight it gave him when Jimmy learned to hold his hands palm to palm and say a few syllables of the Four Vows.

Mrs. Tanahashi was deeply impressed by Senzaki and began Zen study with him. In time, as she prospered as a businesswoman, she provided most of his support. The rest of us would leave contributions on his bookshelves, and after the meetings he would go about the room gathering what he called "fallen leaves." If one of the leaves was a twenty-dollar bill, he would carefully put it away "for Shubin San," Mrs. Tanahashi, to help repay her kindness.

It was Mrs. Tanahashi who read an account in a Japanese journal of Nakagawa Sōen, then a Zen monk in seclusion at Mount Daibosatsu in Yamanashi Prefecture of Japan. Senzaki was impressed by the story of a monk who left "cathedral Zen" behind and sought his own realization in a little hut in the mountains. He wrote to him, and the two monks corresponded for several years. It was arranged that Nakagawa would visit Los Angeles, but the war interfered with these plans, delaying them until 1949.

Meantime, Senzaki, Mrs. Tanahashi, and others in the Japanese community of Los Angeles were interned at the Heart Mountain Relocation Camp in Wyoming. This was disruptive and demoralizing for everyone, but the people who had by now gathered about Senzaki made the best of their situation. They practiced zazen together, recited sūtras, and studied the Dharma. Senzaki's American students were also supportive. Ruth Strout McCandless kept his library, each Japanese volume numbered; when he needed a book, he would request it by number, and she would mail it to him.

After the war Senzaki returned to Los Angeles and was given a room rent free on the top floor of the Miyako Hotel by the owner, a fellow internee. It was here that I met him in December 1947. At this time he had perhaps thirty American students and another thirty followers who had been with him at Heart Mountain. Fifteen to eighteen English-speaking students would crowd into his room for zazen and a lecture two evenings a week. The Japanese students would come for sūtras and zazen on Sunday mornings. We sat on folding chairs, and there was very little ceremony.

We had, moreover, no membership arrangement and no organizational structure. Ironically, this anarchistic arrangement meant that our teacher made all the decisions in a benevolent but authoritarian manner. We were content with this process. He was our wise, gentle teacher, and we could only be followers.

A tolerant teacher as well: Senzaki was not only infinitely patient with us, he welcomed visitors warmly, even Theosophists and spiritualists whose ideas he found ridiculous. I remember one day overhearing his conversation with someone who was holding forth on the occult mysteries of the Pyramids, and his part in the discussion was to say respectfully at intervals, "Oh, really? I didn't know that."

He was especially tolerant of other forms of Buddhism. "Buddhism is a single stream," he would say, and he deliberately used classical terms in their Pali pronunciations, in keeping with Theravada tradition, rather than in the Sanskrit of the Mahayana. For example, he always said (and wrote) "Dhamma" rather than "Dharma." He was friendly with the only Theravada Buddhist teacher in Los Angeles at that time and often invited him to speak to us.

I have never known a teacher more down to earth in his terminology. Once when I was cleaning the dōjō, he was in the library speaking with visitors about his heritage. At one point he raised his voice and said to me, "Please bring me the picture of that old fellow on the table." The old fellow was Bodhidharma, and the table was his altar. Of course Bodhidharma was his inspiration and the table was the devotional center of his practice, but for him these matters were better understated.

Senzaki was a calm and jovial man, at home in the New World, who loved to visit the Sweet Shop in Japan Town with a few friends for a waffle. He walked rapidly, always leading the way, posture very erect for his rather stout figure, with a ready smile and a greeting for his many friends. His clothing was tweedy and rumpled, and for Zen meetings he would simply wear a robe over his street clothes.

In the meetings his talks were full of Zen stories, incomprehensible but delightful. We lived in hopes that we would gradually come to understand them, assured by his words, "Zen is not a puzzle; it cannot be solved by wit. It is spiritual food for those who want to learn what life is and what our mission is."[8] He included personal interviews in the schedule of meetings in early years, but discontinued this practice soon after I began attending.

Senzaki felt he was just introducing Zen practice to the United States. "Some day the Mentorgarten will disappear," he said to me, "but Sōen San will build a great temple in the United States and the Dhamma will flourish." Sōen San, however, became Sōen Rōshi and abbot of a monastery in Japan, and the best he could do was visit.

Like Hotei, however, Senzaki has many descendants. His friend Sōen Rōshi encouraged Mentorgarten students to persevere, and in turn influenced his friend, Yasutani Haku'un Rōshi, to visit the United States and lead retreats during the 1960s. Thus the two rōshis continued Senzaki's work and in time inspired the development of a number of American centers. The Diamond Sangha and its network of centers, the Zen Center of Los Angeles, the Zen Studies Society in New York, and the Rochester Zen Center—all can trace their heritage through the gentle train of karma that Senzaki began. Members of the San Francisco Zen Center and other groups and many individual Zen students feel an affinity with him as well.

After the Miyako Hotel was sold, Senzaki lived out his last years in a flat rented for him by Mrs. Tanahashi in Boyle Heights in East Los Angeles. He continued to meet with students almost to the end and recorded his last words before he died in March 1958. I vividly remember sitting in the funeral parlor and listening to him speak for the last time:

Friends in Dhamma, be satisfied with your own heads. Do not put any false heads above your own. Then minute after minute, watch your steps closely. Always keep your head cold and your feet warm. These are my last words to you.

Then he added, "Thank you very much, everybody, for taking such good care of me for so long. Bye-bye." And the record ended with his little laugh.

I am pleased that *Buddhism and Zen*, the collection of his essays that he assembled in collaboration with Ruth Strout McCandless, is now being reissued. I feel his presence as I read his words:

America has had Zen students in the past, has them in the present, and will have many of them in the future. They mingle easily with so-called worldlings. They play with children, respect kings and beggars, and handle gold and silver as pebbles and stones.[9]

These are words of prophecy, and they are also vows. I make them my own.

Foreword to *Buddhism and Zen*, by Nyogen Senzaki and Ruth Strout McCandless (San Francisco: North Point Press, 1987).

Remembering Sōen Rōshi

When I met the monk Nyogen Senzaki in the Miyako Hotel in Los Angeles at the end of 1947, he told me about Nakagawa Sōen Oshō, with whom he was corresponding, and showed me his picture, a snapshot taken in a field in Manchuria some years before. Though we were just getting acquainted, Senzaki confided in me that he had wanted to bring Sōen Oshō to the United States before the war, but the hostilities had prevented this. Now he was planning again to bring him for a visit. It was clear that my new sensei (teacher) was deeply invested in their friendship, and I got the impression, confirmed many times subsequently, that he hoped Sōen Oshō would eventually settle in the United States as his successor.

I studied with Senzaki Sensei for a few months and returned to Hawai'i before Sōen Oshō visited Los Angeles the following winter. Then in 1950, I received a fellowship to study haiku and Zen in Japan. This involved settling in the Tokyo area, auditing courses at the University of Tokyo, and living in Kitakamakura where I could do zazen and attend sesshin at Enkakuji, the Rinzai monastery where Senzaki Sensei, Professor D. T. Suzuki, and Professor R. H. Blyth had connections. The practice at Enkakuji proved too difficult, so I wrote to Sōen Oshō, explaining that I had been a student of Senzaki Sensei and asking if I could visit him. I enclosed a haiku:

> The train whistle in autumn
> has the same tone
> as the temple bell.

He responded by telegram with the haiku:

> Under red maple leaves
> at our mountain temple
> I am awaiting you.

I found my way to Ryūtakuji and recall vividly our meeting. The monk who met me in the genkan (entryway) had me wait there briefly, and then Sōen Oshō appeared, almost shyly, very young in appearance, although he was then forty-three. "I am Sōen," he said. I had the strong conviction: "This is my teacher."

I tell about moving into Ryūtakuji and practicing there in "Willy-Nilly Zen," published elsewhere,[1] so I will focus here on my memories of Sōen Oshō himself during that period. He became rōshi of Ryūtakuji while I was there, and the events of that spring of 1951 were, I can see now, a kind of forecast of what his life was to be thereafter.

I was thirty-three at this time, full of anxieties about personal matters and full of hopes that a realization experience would put them all to rest. Sōen Rōshi did everything he could to help me to settle down, and though my primary interest was in zazen, he encouraged me to continue to write haiku. I wrote them in Japanese and he offered corrections, and some were even published in a journal. Looking back, I can see that he felt as Bashō did, that haiku is a practice. I did not understand this at the time and did not make full use of him as a poetry teacher, though I wrote perhaps twenty poems during the spring and summer of 1951 when I was in residence at the monastery.

Sōen Rōshi's own great creative period—the last years of the war and the first years of the American occupation—had already passed, but he still wrote occasionally. On one of our excursions together, he composed the haiku:

> The *minomushi*
> has its established place
> among the cherry blossoms.

The *minomushi* is the bagworm, a moth that gestates in little cocoons that hang in cherry trees, and is a favorite subject of Japanese poets. The niche of the bagworm is within the cherry trees, a rather Platonic notion, as in Wallace Stevens's "Anecdote of Men by the Thousand":

> The dress of a woman of Lhasa,
> In its place,
> Is an invisible element of that place
> Made visible.[2]

I remember how the rōshi recited his poem over and over in his magnificent voice—*sang* it, really:

> *Minomushi no*
> *tokoro sademeshi*
> *hana no naka*

"Bashō taught us," he said, "'When you write a poem, you must recite it one hundred times.'" And he would again intone, *"Mino-mushi no / tokoro sademeshi / hana no naka."* And he exclaimed, "Ah, very good!"

Later on, I was to call Sōen Rōshi the Balanchine of Zen because of the way he would choreograph his students and friends into ecstatic bowing exercises or kinhin (walking meditation) through the garden. Now that I understand Balanchine a little better, I realize more than ever how true my words were. Balanchine was deeply conscious of the religious wellsprings of his Russian Orthodox heritage, and his dances were the fulfillment of his spiritual experiences. Sōen Rōshi was intimately in touch with his Buddhist origins, and as an artist of the body, his way was to act them out and to encourage others to act them out as well.

Once when I was visiting Ryūtakuji, a class of perhaps forty junior high school girls came to the temple on a field trip. They were about thirteen years old, shy yet playful, bursting with nervous laughter at the slightest provocation. He organized them into a za-zenkai in a matter of minutes and had them doing zazen and kinhin with serious demeanor. Then he let up and served informal tea and answered questions. My impression is that later, they would always look back on this experience as a milestone in their growing up.

I could also compare Sōen Rōshi to Black Elk. When Black Elk had a vision, his people would act it out as a pageant. If there were horses of a particular color and marking in his dream, horses of that sort were found for the pageant. The costumes were the same, the words were the same, all the movements were the same as in his dream. His visions were thus confirmed in the world, just as Balanchine's were. (I have always thought that the book *Black Elk Speaks* could be turned into a marvelous ballet.)

Sōen Rōshi was sometimes low key in the enactment of his dreams—in tea ceremony, for example. Before I went to Ryūtakuji I had never seen a tea ceremony. I took tea with him almost every morning, but it was not until we traveled to Kyoto together and I saw what I knew to be tea ceremony that I realized I had been participating in the same ceremony with him, there in his room as the birds began singing in the early mornings. Those occasions were profoundly enjoyable, but I had no idea that we were following the pattern of an old ritual.

"Every act is a rite," Thich Nhat Hanh has said. I am sorry the two teachers never met, for they would know each other at first glance. "When you sweep the garden, you are sweeping your own mind," the rōshi said to me. I felt that he was repeating a rather obvious old Zenish expression. I did not appreciate how it was possible to personalize such a rite. I did not understand how samu (work practice), as set forth by old Pai-chang and his predecessors, really is the act of Shākyamuni Buddha turning the Dharma wheel.

I realize, however, that Sōen Rōshi accepted me the way he did because he felt that I understood, to some small degree, his ritualistic imperatives. When we visited Sen no Rikyū's teahouse in Kyoto, he threw back the covers protecting the precious old tatami and mimed a tea ceremony with me. I entered into the game fully and could even do a little mondō (Zen dialogue) with him.

I don't pretend to know, across the cultural and language barriers as well as the lay-clerical barrier I felt at Ryūtakuji, just how his old teacher, Yamamoto Gempō Rōshi, and the monks at the monastery felt about the powerful ceremonial urges that motivated everything that Sōen Rōshi did from the moment he arose in the morning to his act of lying down at night. I would guess that they did not understand him completely and that many of the monks simply humored him, shrugging at what they considered to be strange behavior.

People in the wider Zen community also seemed to have doubts about him. I remember Professor Suzuki cautioning me about Sōen Rōshi in the summer of 1951. "You know," he said, "he's a rather peculiar fellow. After he was ordained, he went off and lived by himself at Daibosatsu Yama in Yamanashi Prefecture for two years with very little contact with his teacher or the other monks at Ryūtakuji." By that time, I had visited the rōshi's old retreat at Daibosatsu Yama and had seen some of the ceremonies he had held there by himself—for example, building cairns while singing an old folk song about remembering one's parents. Yes, he was peculiar, in the first meaning of the word: "distinctive, unique"—and Zen, in Japan at any rate, is a religion with rather exacting conventions.

One of those conventions was the Shinzanshiki, the ceremony in April 1951 that installed him as abbot of Ryūtakuji. That spring he was very busy preparing the many necessary details for the occasion. Following his installation we had our first sesshin with him, while Gempō Rōshi set out to enjoy his retirement. I remember that our new rōshi gave teishō (a talk on the Dharma) on the *Rinzai Roku*, but his voice was so soft and his manner so diffident that it did not seem like teishō at all. He sat on a cushion facing the altar, rather than on the high seat where Gempō Rōshi had sat before.

Part of this was Japanese modesty, I suppose, not wishing to assume a new responsibility too quickly, but part was surely genuine reluctance to take on the role of master. Perhaps some people might have supposed that he felt he was not adequately prepared for the role, not far enough along in his practice, but I would conjecture that he simply considered himself quite unsuited, although his teacher believed that he was just the right successor. And it was his loyalty to his teacher that came first.

After the sesshin, he secluded himself at Daibosatsu Yama, and Gempō Rōshi had to return to lead the May sesshin. I recall that Gempō Rōshi devoted parts of each teishō at that time to excusing Sōen Rōshi, explaining that he had tired himself out in preparation for the Shinzanshiki and assuring us that our new rōshi would be all right.

Was his confidence misplaced? Sōen Rōshi told me many times that he could not be my teacher, and I have been told by others that he told them the same thing. He was a little like Krishnamurti, who for very different reasons has had difficulty accepting his role as a teacher. At the same time, he often said to me that it was the first duty of a rōshi to find a successor with clear eyes. "If he does not have two clear eyes, at least he should have one clear eye. If not one eye, at least half an eye." I believe that he, in his great modesty, considered that he might have half an eye and that it was his karma and his responsibility to accept his position as wholeheartedly as possible and to maintain and pass on the teaching that he had received from the Buddha through Bodhidharma, Hakuin, and Gempō.

He could not keep all his commitments. After the deaths of his mother and of his old teacher, Gempō Rōshi, he stopped coming to the United States for sesshin with students of Senzaki Sensei, who had died in 1958. In 1962, he referred those students to Yasutani Haku'un Rōshi, who continued trips for sesshin to this country until 1969. It was clear to us that Sōen Rōshi could not live up to Senzaki's hopes that he would settle in the United States. It is apparent now that he felt that the monk Eidō Shimano would fulfill those expectations in his place.

The other monks at Ryūtakuji seemed to have doubts about Shimano. I remember my surprise when I visited the monastery in 1961, a year after Shimano had come to Hawai'i as our resident leader. The Ryūtakuji monks were my old friends by then—we had known each other long before they had known Shimano. I was struck by the fact that none of them inquired about him, even though he had left the monastery only the year before and I had just come from living with him. I asked one of the monks about this and received a look and a shake of the head that clearly informed me that Shimano was not one they could accept.

I am sure that they found ways to communicate their doubts to Sōen Rōshi. Why should he ignore the opinions of his students? Perhaps he felt that his own path was one that was more suited to the United States, a fresh environment, than to the old tradition-bound monastery setting in Japan. When Sōen Rōshi became convinced that Shimano understood this particular path of Zen, then perhaps the rōshi was also convinced that he had found the successor he had sought, one who could follow in the footsteps of Senzaki Sensei. Perhaps he also felt that the monks could not understand this, since they had not understood him in other ways.

When Shimano's social relationships got him into trouble at the Koko An Zendō in Hawai'i in 1964, he felt obliged to move to New York. My own relationship with Sōen Rōshi fell apart at this point. He could not believe that Shimano's behavior was not just that of a "young rascal." Although we saw each other from time to time after that and remained on fairly cordial terms, I always felt that Sōen Rōshi blamed me to some extent for Shimano's failure to keep his commitments in Hawai'i.

Sōen Rōshi continued to be faithful to Eidō Shimano over the years that followed. His initial belief that Shimano understood him and his imperative to find a successor with at least half an eye apparently kept his confidence unshaken until the crisis created by allegations of sexual abuse at the New York Zen Studies Society erupted in 1975. Moreover, his confidence seemed renewed to some limited degree from time to time even after that, almost until the end of his life. The reasons for this continued support have not always been clear to Zen students in the New York sangha and elsewhere in this country.

They are not completely clear to me either, but as best I can understand them, it seems that three factors are involved. The first is that Japanese social relationships are established on Confucian standards of loyalty to the superior and responsibility for the inferior. Translated into terms of teacher and student, this means that the Japanese Zen student and teacher support each other instinctively as part of their cultural mores.

The second factor is that the student and teacher create a special bond over the years of this intimate, one-on-one interaction in the dokusan (interview) room and in their day-by-day association in life together in the monastery. Their affection for each other is as deep as can be found in any family.

Sōen Rōshi's loyalty to his disciple under trying circumstances can be compared with the action of his ancestor, Tōrei Zenji, who, I have heard, disowned and defrocked his successor for a major violation of trust in connection with the rebuilding of Ryūtakuji after a fire. I sense that Tōrei felt the monk had violated the Dharma and that this betrayal was serious enough for him to set aside the tradition of personal loyalty to one's student. Otherwise, Tōrei's own teaching of accepting abuse could be called into question.[3]

The third factor is Sōen Rōshi's own personality. He was profoundly faithful by nature. In his earliest days at Ryūtakuji, he installed his widowed mother in a cottage on the compound of the monastery, where she remained until her death many years later. He called upon her almost every day when he was in residence and read her his mail and listened to her comments. His relationship with Gempō Rōshi was that of an adoring son, and when the old teacher died, he mounted his life-sized photograph in the main hall of the monastery. It dominated the room, while the Kanzeon figure on the altar reposed behind its screen.

Once, Sōen Rōshi asked me, "What is the most important thing in the world?" I did not dare to answer, so he replied for me, "I think friends are the most important thing in the world." Shimano was much more than a friend, and I can only guess at the deep despair he must have felt when he could finally acknowledge to himself that he had been gravely mistaken about him.

From afar, I always wondered if Sōen Rōshi's extended private retreats in his later years were related to a sense of betrayal by Shimano. His life and his commitments must have seemed to him to be unfulfilled. All the time I knew him, zazen was the way he restored himself, and he believed in the power of one's own zazen to restore others. But zazen in retreat cannot influence others unless they are open to influence. I mourn our great teacher and the tragedy of those final years.

1984

Remembering Blyth Sensei

I was a civilian internee from Guam, held with forty-four other men in what we called the "Marks House" (named after the former British owner) in the foreign district of Kobe, near the Tor Hotel. By the winter of 1942–43, we had been interned for one year in Japan. I was not too well, suffering periodic bouts of bronchitis and asthma, but I kept up with my reading from the library we had brought with us from our first camp in Kobe, the Seaman's Mission on Ito Machi. I also had bought books with money supplied to us through the Swiss consul by the United States government. These books included works on haiku in English, a subject that had interested me before the war.

One evening a guard came into my room, quite drunk, waving a book in the air and saying in English, "This book, my English teacher . . ." He had been a student of R. H. Blyth at Kanazawa, and the book was *Zen in English Literature and Oriental Classics*, then just published.[1] I was in bed but jumped up to look at the book and was immediately fascinated. I persuaded the guard to lend it to me, and weeks later he bought another copy for me so that he could have his own copy back.

I suppose I read the book ten or eleven times straight through. As soon as I finished it, I would start it again. I had it almost memorized and could turn immediately to any particular passage. It was my "first book," the way *Walden* was the "first book" for some of my friends, the way *The Kingdom of God Is Within You* was the "first book" for Gandhi. Now when I look at it, even the type looks different, far smaller in size, and the references to Zen seem less profound. But it set my life on the course I still maintain, and I trace my orientation to culture—to literature, rhetoric, art, and music—to that single book.

In May 1944, all the camps in Kobe were combined, and we were housed in a former reform school called Rinkangaku, in Futatabi Park near Nunobiki Falls in the hills above Kobe. Mr. Blyth had been held in one of the other camps, and now at last we could meet. I think he was rather overcome by my adulation, and he rejected it at first, not wanting, as he said, a disciple. But we straightened out this initial misunderstanding and soon established the intimate relationship we were to maintain until his death in 1964, and which we still maintain today, though he has been dead a long time.

Rinkangaku consisted of three large, connected buildings, containing dormitories, commons rooms, and classrooms. One hundred seventy-five men completely filled this complex, and Mr. Blyth lived with six others in what had apparently been the commons room for the teachers. He had his bed in the *tokonoma*, the alcove usually reserved for scroll and flower arrangement in Japanese homes and offices. His books teetered on shelves he himself had installed above the bed. All day long he sat on his bed, sometimes cross-legged and sometimes with his feet on the floor, writing on a lectern placed on a bedside table, with his reference books and notebooks among the bedcovers. It was during this time that he was working on his four-volume *Haiku* and his *Senryu*, as well as *Buddhist Sermons from Christian Texts*[2] and other works. I recall that he wrote rapidly, with his words connected, using two sets of pen and ink, black for his text and red for his quotations.

With my interest in haiku and my new enthusiasm for Zen, we agreed that I should learn Japanese, so he obtained some elementary school texts from his Japanese wife (families were permitted a weekly visit) and gave me a couple of dictionaries. He loaned me what Zen texts he had; I remember particularly D. T. Suzuki's *Essays in Zen Buddhism: First Series* in the original Luzac edition and Wong Mou-lam's *Sutra of Wei-lang* in its original paperbound edition, from Shanghai, I believe.[3]

I would study and read during the day while he wrote at his lectern, and in the evening I would visit his room to read my lesson to him, perhaps show him a new haiku I had written, and talk generally with him and with his roommates, all of them old Japan hands, some of them partly of Japanese ancestry.

Generally, Blyth Sensei was well liked by his fellow internees, and though he was regarded as pro-Japanese by the Americans from Guam, they respected his learning and his diligence and knew that they could get straight talk from him and learn something in the process. They called him Mr. B., a nickname he rather liked. It seemed to express the balance between familiarity and formality that he sought.

With the part-Japanese internees, he seemed to have a more uncertain relationship. Perhaps they regarded him as a kind of Johnny-come-lately. They busied themselves with studies of Japanese politics, history, and economics, while he devoted himself to literature and religion. He would always swing the talk from their interests to his own, which he regarded as more fundamental. They seemed to feel that his interests were quite outdated and irrelevant, and there were many heated arguments in his crowded, smoky room.

If the American internees had known their Mr. B. more intimately, they would have understood that his attitude toward Japan was realistic and not blindly supportive. He had begun the process of applying for Japanese citizenship before the war, but he allowed this process to lapse after the war began, saying that if Japan lost the war, then he would renew his application. (It turned out that he died a British subject.)

"Can you imagine people like these guards occupying your country?" he once asked me. Somehow he sensed how badly the Japanese were handling their responsibilities as occupation forces in Southeast Asia, and he felt that a national defeat might be the salutary experience the country would need for true maturity.

So my lessons from Blyth Sensei included political science, as well as language, literature, and religion. He seemed to regard his understanding of culture as the ground for making judgments. After the war, while teaching and in residence at the Gakushūin (the Peers School), he pointed to an encyclopedia in his bookcase and said, "I would like to make a book of commentary that would follow the main topics of that compendium of facts."

Blyth Sensei often mentioned how as a boy he was inspired by Matthew Arnold's ideal of developing one's self to the fullest, to be one's own best linguist, musician, artist, and scholar. Thus he learned Spanish in order to read *Don Quixote*, Italian to read Dante, German to read Goethe, and he made a valiant effort to learn Russian in order to read Dostoyevsky. And, of course, he was a deep student of Chinese, Korean, and Japanese.

As a musician, he loved the oboe particularly but played virtually all the Western orchestral instruments. While he was at the Gakushūin, he constructed an organ, a remarkable feat of technical and musical skill. His words about Bach influenced my taste during the war and directed me on the path of music appreciation I still follow. And somehow just his passing words on Turner, Sesshū, and other artists established my understanding of art.

There were flaws in this Renaissance man, however. He did not go far enough in his Zen practice to justify his confidence in commenting on the *Mumonkan* (*The Gateless Barrier*) and that book is probably the weakest of his works.[4] He loved women and scorned them, his relations with those close to him were stormy, and his remarks about women, particularly in the essays he published after the war, infuriate readers and alienate them to this day.

I accept these flaws as I accept the flaws in my own father. The one brought me into physical being and shaped my character, the other put me in touch with myself and with this rich, wonderful world. If we had not met, I might well have spent my life mundanely, saying and doing trivial things. His words rise in my mind as I speak to my own students, and his face still appears in my dreams.

Translated in *Kaisō no Buraisu* [*Recollecting Blyth*], edited by Shinki Masanosuke (Tokyo: Kaisō no Buraisu Kankōkai Jimushō, 1984).

Openness and Engagement

Memories of Dr. D. T. Suzuki

I FIRST encountered Dr. Suzuki's name in R. H. Blyth's *Zen in English Literature and Oriental Classics*, which I read in an internment camp in Kobe, Japan, in the winter of 1942–43. Later on when our camps were combined, I met Professor Blyth in person, and he told me about his first conversation with Suzuki Sensei:

> BLYTH: I have just come from Korea, where I studied Zen with Kayama Taigi Rōshi of Myōshinji Betsuin.
> SUZUKI: Is that so? Tell me, what is Zen?
> BLYTH: As I understand it, there is no such thing.
> SUZUKI: I can see you know something of Zen.

If there was challenge in Sensei's words, it was of the mildest sort. His fundamental purpose was to encourage. Many scholars and students of Zen can tell similar stories—I think especially of Richard DeMartino, Philip Kapleau, and Chang Chung-yüan.

My own first meeting with Sensei was in 1949 at the Second East-West Philosophers' Conference at the University of Hawai'i. That was a wonderful summer. There were many stars at the conference, particularly from India, but Sensei by his manner (for few could understand him) stole the show. It was just after F. S. C. Northrop had published *The Meeting of East and West*.[1] Everyone was uncomfortable with the conceptual formulations in this work, but only Sensei could pinpoint the problem. I remember the chuckles of amusement among the scholars when he remarked, "The trouble with the 'undifferentiated aesthetic continuum' is that it's too differentiated."

I was part of a clique of graduate students who attached themselves to Sensei, and we attended (or crashed) many dinners and receptions that were given for him by University of Hawai'i dignitaries and by Japanese American organizations in the Honolulu community. Richard DeMartino was his secretary and companion at that time and had purchased a Model A Ford for their transportation. Those were the days when the Model A was just an old car, not a precious antique, and I remember the endearing sight of Sensei rattling up to distinguished gatherings in that aged clunker, full of dignity and good humor. I wanted to continue my study of Zen and asked Sensei's advice: "Should I return to Los Angeles and study with Nyogen Senzaki, or should I go to Japan?" "Go to Japan," he said, and he wrote the letters I needed for my visa.

In the summer of 1951, I called at the Matsugaoka Library in Kitakamakura, where Sensei had just returned from his two years in the United States. I was ill from the rigors of monastic living, and Sensei insisted that I stay with him and recuperate. I remained with him for two weeks, as I recall. Sensei saw that I was well cared for by his staff, and he included me in all of the gatherings at the library; I remember particularly a memorial service for Beatrice Lane Suzuki, attended by his old friends.

We had many conversations about religion in the United States. Sensei had been inspired by his experiences at Columbia University, and I wish now that I had kept a record of his words about the scholars he had met. I do recall him saying that he felt more affinity with anthropologists than with Protestant theologians.

Thereafter, down through the years until his death, we kept in touch. When he visited Hawai'i or when Anne Aitken and I visited Japan, we always had tea or a meal together. Anne recalls a dinner given for him by the Young Buddhist Association of Honolulu in 1959. We were standing around afterward, waiting for the dishes to be cleared away, and she noticed Sensei browsing among the tables, picking parsley off the plates and eating it. Catching her eye, he grinned like a little boy and said, "People don't eat their parsley, and it is so good for them." She was moved by his sensitive expression of responsibility for others, including the parsley that would otherwise be sacrificed for nothing.

The most memorable of those later meetings took place during his last trip to Hawai'i in the summer of 1964. He spoke to a packed house at the Koko An Zendō, and in the question period, a student asked, "Is zazen necessary?"

Sensei replied, "Zazen is absolutely not necessary." This created quite a stir among the Koko An members.

The next year, Professor Masao Abe visited the East-West Center, where I was on the staff. We met on the steps of Jefferson Hall and greeted each other.

"Mr. Aitken," said Professor Abe, "I hear that Suzuki Sensei spoke at Koko An last year. Is that correct?"

"Yes," I said, "it is."

"I hear," continued Professor Abe, "that he said zazen is not necessary. Is that correct?"

"Yes," I replied, "he said zazen is absolutely not necessary."

"Oh," said Professor Abe, "he meant zazen is relatively necessary."

Now that was very clever of Professor Abe, and it served to highlight Sensei's unorthodoxy. He knew very well, but seldom said, that zazen is relatively necessary. He was, however, critical of Alan Watts and others who dismissed zazen as unimportant.

Comparing notes about our old teacher, Anne Aitken and I find that we both asked him, at different times, about the interpretation that Mr. Watts gave to a story about Nan-yüeh and Ma-tsu. Nan-yüeh found Ma-tsu doing zazen and asked him what he was trying to do. Ma-tsu said that he was trying to become a Buddha. Nan-yüeh thereupon picked up a piece of roofing tile and began rubbing it with a stone. When Ma-tsu asked him what he was doing, Nan-yüeh said he was making a mirror out of the tile. "No matter how you rub that tile," said Ma-tsu, "it will never become a mirror." Nan-yüeh replied, "No matter how much you do zazen, you will never become a Buddha."[2]

Mr. Watts remarks somewhere in his books that this dialogue

showed how T'ang period Zen people disapproved of zazen. Dr. Su-zuki said to both Anne and me, "I regret to say that Mr. Watts did not understand that story."

Still, Sensei hardly ever mentioned zazen in his writings. Even *The Training of the Zen Buddhist Monk* barely touches on this funda-mental aspect of Zen life.[3] Now that I am more intimately involved in Zen practice, I would like to talk with him about zazen and other matters. It would be a long conversation. I would want to take up the nature of the kōan, the place of prajñā and the mind, the function of words, and the writings of Dōgen Zenji.

He would listen—he always did. Once, in a class at the Univer-sity of Hawai'i, I asked him about a version of Chiyo-ni's haiku, "The Morning Glory," that he had written on the blackboard. In transliterated Japanese, this verse reads:

> *Asagao ni*
> *tsurube torarete*
> *morai mizu*

It is usually translated:

> The morning glory
> has taken the well bucket;
> I must ask elsewhere for water.[4]

However, Sensei had written "*Asagao ya . . .*" on the blackboard. The substitution of *ya*, a cutting word that might best be translated with a colon or an exclamation mark, for *ni*, a postposition indi-cating an agent, would make the translation:

> The morning glory!
> It has taken the well bucket;
> I must ask elsewhere for water.

This changes a rather precious poem about someone who finds that the morning glory has entwined the bucket and does not want to disturb it into a Zenlike poem about someone who is struck by the beauty of the morning glory and can only exclaim, "The morning glory!"—and then, as an afterthought, considers borrowing water from the neighbors.

Anyway, I knew the conventional version, and I suspected that Sensei with his Zen attitude had inadvertently imposed his own revision. He listened to me, and wrote to scholars in Japan and learned that indeed there was some speculation that *ya* was the original particle in the first line. He did not stop there but went on to discuss the matter in class and then to write his cogent essay, "The Morning Glory."[5] Incidentally, this essay contains Sensei's clearest presentation of a concern that preoccupied him during his later years— world peace. Clearly, he felt that people are not sensitive to flowers or to the sacrifice of parsley, and so we have nations threatening each other with nuclear weapons.

The development of that essay is an example of Sensei's creative process generally—openness and engagement. He would listen or read with an open mind, and then involve himself in considering the matter, and finally come forth with his own unique response.

Openness and engagement show in his face in these sensitive portraits by Francis Haar, the purity and wisdom of a very old man who has devoted his life to the Tao. They evoke his inspiring presence and remind me that I need not wait for some kind of miraculous logistical arrangement for our conversation.

"Now, about the importance of zazen, Sensei . . ."

First published in *A Zen Life: D. T. Suzuki Remembered*, photographs by Francis Haar, edited by Masao Abe (Tokyo and New York: Weatherhill, 1986).

The Legacy of Dwight Goddard

Most intellectuals can look back to a "first book" that gave coherence to their interests and set them on their life's course. Students of Zen Buddhism come to me with a variety of "first books" in their past, and among them, with some frequency, is Dwight Goddard's durable anthology of translations, *A Buddhist Bible*, originally published in 1932 and then republished in its present enlarged form in 1938.

As the "first book" for Jack Kerouac, *A Buddhist Bible* had a direct influence on the American Beat movement of the 1950s—and thus on the New Age movement that followed, with its efflorescence of Western Zen Buddhism, in the late 1960s and early 1970s. Allen Ginsberg wrote of Kerouac:

> He went to the library in San Jose, California, and read a book
> called *A Buddhist Bible*, edited by Dwight Goddard—a very
> good anthology of Buddhist texts. Kerouac read them very
> deeply, memorized many of them, and then went on to do
> other reading and other research and actually became a brilliant, intuitive Buddhist scholar. . . . He introduced me to
> [Buddhism] in the form of letters reminding me that suffering
> was the basis of existence, which is the first Noble Truth in
> Buddhism.[1]

In *Jack's Book*, Barry Gifford and Lawrence Lee expanded on the importance of *A Buddhist Bible* for Kerouac:

In its 700-odd pages he found concepts of historical cycles so gigantic that they dwarfed Spengler's. He found, as well, the notion of *dharma*, the same self-regulating principle of the universe that he had proposed in the closing pages of *Doctor Sax....* Using his sketching technique Jack converted the texts in *A Buddhist Bible* into his own words.[2]

This "translation" began a creative process of Americanizing Buddhism that was manifested first in Kerouac's *San Francisco Blues* and flowered in *The Dharma Bums*,[3] which itself became a "first book" for people growing up during the 1960s. In that era, I met many people whose ruminations echoed those of Ray (Kerouac himself) and Japhy (Gary Snyder). Here is Ray, looking around at his friends sleeping in the early morning, for all the world like the young Gautama viewing his sleeping retainers before he set out on his lifetime pilgrimage.

> I suddenly had the most tremendous feeling of the pitifulness of human beings, whatever they were, their faces, pained mouths, personalities, attempts to be gay, little petulances, feelings of loss, their dull and empty witticisms so soon forgotten. Ah, for what? I knew that the sound of silence was everywhere and therefore everywhere was silence. Suppose we suddenly wake up and see what we thought to be this and that ain't this or that at all? I staggered up the hill, greeted by birds, and looked at the huddled sleeping figures on the floor. Who were all these strange ghosts rooted to the silly little adventure of earth with me? and who was I? Poor Japhy, at eight a.m. he got up and banged on his frying pan and chanted the "Gacchami" chant and called everybody to pancakes.[4]

Jack Kerouac, cross-fertilizing with Snyder, Ginsberg, Philip Whalen, and others who are still engaged in Americanizing Buddhism in their own ways, helped to establish a culture in which the San Francisco Zen Center could grow and flourish in the mid 1960s. A number of Zen Buddhist centers in San Francisco and Berkeley have emerged in the generation that has followed. When I visit and give a public talk in one of those cities today, I find that I can, without watering anything down, use the same Sanskrit and Japanese terms and Buddhist concepts that I do in classes with my own students. Everyone is following along and even getting ahead of me. The Bay Area is Buddha Land, and there are similar Buddha Lands, less obvious perhaps, across the country and across the Western Hemisphere.

A Buddhist Bible was an important seed in this acculturation. The book was composed, as Goddard states in his preface to the 1932 edition, to record adaptations of the original teachings of the Buddha, from the rise of the Mahayana to the development of Dhyāna Buddhism to the *Platform Sūtra* of Hui-neng, the Sixth Ancestor of Ch'an or Zen Buddhism. Goddard remarks that Buddhism "is the most promising of all the great religions to meet the problems of European civilization which to thinking people are increasingly foreboding." He felt that Japanese Zen was "the purest form of Buddhism and the closest to the teachings of Gautama Buddha, its founder."[5] This was a privately expressed view, however, for *A Buddhist Bible* is a broadly inclusive anthology that serves Buddhist readers generally. It brings together a collection of sacred texts that are available in more up-to-date translations elsewhere, perhaps, but are scattered in various publications, some of them out of print.

It is tempting to compare *A Buddhist Bible* with D. T. Suzuki's *Manual of Zen Buddhism*, published in the same decade.[6] Indeed, the two books overlap to some extent, for both contain the *Heart Sūtra* and selections from the *Diamond, Shūrangama, Lankavatara,* and *Platform* Sūtras. Suzuki and Goddard selected different sections from the original texts, however, and their translations vary. The *Manual of Zen Buddhism* is quite sectarian, whereas *A Buddhist Bible* includes a broad selection of Mahayana texts as well as selections from the Pali and Tibetan canons.

Goddard honored the traditional Chinese classification of Mahayana Buddhist texts, which places their origins in India for the most part. Research into the origins of Buddhist literature is ongoing today, but even in Goddard's time it was clear to Western scholars that many texts attributed by the Chinese to Sanskrit sources were in fact probably written in China in the fourth through seventh centuries. In Goddard's "Selections from Sanskrit Sources," the *Shūran-*

gama Sūtra, *The Awakening of Faith*, and probably the *Lankavatara Sūtra* fall into this uncertain category.

In other respects, however, Goddard felt free of tradition. In his preface to the 1938 edition he discusses his rationale for cutting "a great deal of matter not bearing on the theme of the particular Scripture" and also his readiness to interpret "where it seemed necessary and advisable." He also was quite willing to edit the English of his translators, including D. T. Suzuki, and to move paragraphs and sections around within a traditional text—thus showing the confidence and courage of a talented autodidact unhampered by scholarly constraints.

The *Diamond Sūtra* section is a prime example of his editing. When Goddard spoke of texts that "are often overloaded with interpolations and extensions . . . in places confused and obscure," he was surely referring to the *Diamond Sūtra* in particular. Working from the translation provided by his associate, the monk Wai-tao, he forthrightly rearranged this sūtra by the Six Pāramitās—the traditional "perfections" of giving, morality, patience, zeal, focused meditation, and wisdom. On examining the original, I find references to "giving," for example, scattered in six of the *Diamond Sūtra*'s thirty-two short chapters. Goddard assembled these widely separated references into a single section entitled "The Practice of Charity." He followed the same process when assembling the other five sections of the sūtra by the pāramitās, giving it a kind of coherence it lacks in more faithful translations.

It is a problematic coherence, of course. Scholarly constraints keep translations as true as possible to the original intentions of a text. As early as the fourth century, Vasubandhu and other Indian Buddhist philosophers suggested various thematic orderings of the *Diamond Sūtra*, but they did not try to rearrange it accordingly.[7] In deference to tradition, Goddard noted the original locations of the paragraphs. Then, by cutting the repetitions and what he considered to be superfluous material, he produced a version that is about two-thirds the length of other English translations. It is not, however, the *Diamond Sūtra* studied, memorized, and chanted by fifty generations of Mahayana Buddhists.

In "Selections from Chinese Sources," Goddard included the Tao-te-ching, the central book of Taoism, as a text that contributed to Ch'an. It is "not strictly a Buddhist text," he confesses. Indeed. But it is a text that profoundly influenced the development of Chinese Buddhism and has been studied from the beginning by Chinese Buddhist monks. In fact, David Chappell, professor of Chinese Buddhism at the University of Hawai'i, remarked to me that the original classical pieces that are translated in *A Buddhist Bible* actually form the syllabus studied by Chinese Buddhist monks over a period of 1,500 years.

Goddard also included a translation from the T'ien-t'ai (Japanese: Tendai) tradition, which he titles "Dhyana for Beginners," a work ascribed to Chih-chi (Chih-i, 538–97), the founder of T'ien-t'ai. This is a tradition that is little known among Western Buddhists, though it was the mother of the sects and subsects that emerged in the Kamakura period of Japanese history (1186–1333): Pure Land, Zen, and Nichiren. It has served as a bridge between Theravada and Mahayana schools as well. Its central term, *chih-kuan* (Japanese: *shikan*) derives etymologically from the two qualities of Theravada practice, *samatha* and *vipassana*, "stillness" and "insight." Goddard explicates the term *chih-kuan* as "dhyāna" and translates it "serenity and wisdom."

The meditation practice of dhyāna is carried on to this day in all Zen Buddhist centers, East and West. It is an oral tradition, however. Aside from Dōgen Kigen's *Fukan Zazengi*, a work probably unknown to Goddard, the practice itself is not set forth in much detail in Buddhist literature. It seems that Goddard intended "Dhyana for Beginners" to fill that gap.

In addition, "dhyāna" traditionally meant far more than stillness and insight. It included personal discipline in daily life and precepts of ethical and harmonious living, topics that underlie Zen Buddhist texts but again are not spelled out discursively. It is interesting to compare words of "Dhyana for Beginners" with kōans of Zen. Here is Chih-chi:

> Our body is very sensitive to softness, smoothness, warmth in
> winter, coolness in summer, etc. We are so ignorant as to the

true nature of these sensations that our minds become upset and foolish by the touch of pleasant things, and our effort to attain enlightenment is obstructed and hindered.[8]

And here is Tung-shan (807–69), venerated as the founder of the Ts'ao-tung (Sōtō) school of Zen:

A monk asked, "How does one escape hot and cold?"

"Why not go where there is neither hot nor cold?" said the Master.

"What sort of place is neither hot nor cold?" asked the monk.

"When it is cold, you freeze to death. When it is hot, you swelter to death."[9]

Chih-chi discourses on ignoring comfort and discomfort, Tung-shan announces the way to do it. Chih-chi points to a step-by-step practice toward forgetting the self; Tung-shan points to the experience. "Dhyana for Beginners," then, offers the first of many steps. And though it was originally composed for monks and might be considered moralistic by Western readers, it nonetheless presents ideals of personal purity that we can translate for our own instruction.

Also in "Selections from Chinese Sources" is the "Sutra Spoken by the Sixth Patriarch" (*The Platform Sūtra*). This text is familiar to Western students of Zen Buddhism, for it is part of *The Sutra of Hui-neng*, published with A. F. Price's *The Diamond Sutra*.[10] Even the translator is the same, Wong Mou-lam, a colleague and friend of Dwight Goddard. This is, of course, the Sung-period edition of the sūtra, not the much earlier version discovered in the Tun-huang caves, which was translated in 1967 with a long historical commentary by Philip B. Yampolsky in his *The Platform Sūtra of the Sixth Patriarch*. Again, Goddard's cutting and rearranging give the sūtra a certain coherence that it lacks in the original.

Goddard did not intend *A Buddhist Bible* to be a sourcebook for critical and literary study. In a letter to Ruth Everett he said about another of his publications, "Whatever I do will inevitably be done in an amateurish way and will have to be redone later by abler minds, but I feel that because of the present situation I must do the best I can, and remain willing to be forgotten by the greater writers who are to follow."[11]

Goddard's "best" remains a building block of Western Buddhist practice, while he himself, though a prolific writer and pamphleteer, is almost forgotten. In Charles S. Prebish's biographical sketch of Goddard for the forthcoming *American National Biography* (Oxford University Press), Prebish remarks at the outset, "The details of his family background and early life are rather obscure." The story of his career is also incomplete. A few letters were kept in the family. His sister, the family genealogist of her time, began a chronology of his life, and this was extended with brief comments by a niece, Alice M. Brannon, who kept house for him for lengthy periods. In addition, Goddard's publications themselves provide some biographical information, a few articles have been written about him, and the First Zen Institute of New York has kept his letters to Ruth Everett and Shigetsu Sasaki. Still, the record is meager and there are many gaps.

Dwight Goddard was born July 5, 1861, in Worcester, Massachusetts. After graduating with honors from Worcester Polytechnic Institute, he began a career in industry. In 1889, when he was twenty-eight, he married Harriet Webber. It was a happy marriage, according to Alice Brannon, but Harriet died just the next year. This tragedy, one can assume, brought Goddard face-to-face with the deepest human questions. The following year, in 1891, he entered Hartford Theological Seminary, where he was graduated in 1894, at the age of thirty-three.

Ordained and posted to China as a Congregational missionary, Goddard attained some status when he was chosen to write the *Report of the Jubilee Year of the Foochow Mission of American Board of Commissioners for Foreign Missions* in 1897. He married Dr. Frances Nieberg, a fellow missionary, and their first son was born in Foochow.

On the basis of interviews conducted late in Goddard's life, David Starry wrote of Goddard's dissatisfaction during those early years in China. "During his initial years as a . . . missionary in southern China, he became increasingly frustrated at the failure of the Christian missions to accomplish their spiritual goals. He was convinced that although the Christian propaganda had been successful in influencing educational and social conditions it had failed in its purely religious aspects."[12] He prowled around with an open mind, visiting Buddhist temples—alert for spiritual nourishment.

Returning with his wife and child to the United States about 1899, Goddard accepted pastoral positions in Lancaster, Massachusetts, and in Chicago. A second son was born. Then abruptly he changed course again, returning to industry as a mechanical engineer. In his spare time he set about writing a series of biographical essays that were collected and published in 1906 as *Eminent Engineers*.

In the course of his engineering enterprise, Goddard sold an invention to the U.S. government that was later used during World War I. This brought him a fortune that supported his family and allowed him to retire from industry in 1913 and resume his religious quest. It was not a simple transition. He lived alone for a while and suffered a nervous breakdown. David Starry conjectures that he was burdened by the thought of his invention being used for purposes of war. In any case, during the next few years "Uncle Dwight was a wanderer," as Brannon noted in her chronology. He lived for periods of a few years in Thetford, Vermont, and in Ann Arbor, Michigan—more briefly in Lancaster and in Los Gatos, California. He became interested in Taoism during this time, while he was also reading and writing in the field of Christian mysticism.

He made several trips back to China. In 1921 he learned about Karl Ludwig Reichelt, a Lutheran pastor who had established a monastery in Nanking that was devoted to Christian-Buddhist understanding. He spent some time in Reichelt's monastery in 1923 and again in 1925.

One is not surprised to learn that with all this wandering, he and his wife were divorced in 1926. He married for a third time a year later, at age sixty-six, and this marriage ended fairly soon afterward in a separation.

In 1928, at the age of sixty-seven, Goddard encountered Japanese Zen Buddhism for the first time through Junsaburo Iwami of New York City. Iwami was at the time attending lectures by Shigetsu Sasaki (later Sokei-an Oshō) at the Orientalia Bookshop. Sasaki recalled that Iwami "got one of Goddard's circulars he was always sending around and wrote to him about Zen Buddhism. Goddard was terribly moved that he never knew Zen Buddhism."[13] After he and Iwami met, Goddard went forthwith to Japan, where he consulted with D. T. Suzuki and studied eight months with Yamazaki Taiko Rōshi of Shokoku Monastery in Kyoto, living apart from the monastery but visiting for zazen and personal interviews. He dedicated the first edition of *A Buddhist Bible* to Suzuki and Yamazaki as his teachers.

In letters to Ruth Everett, Goddard reported on his difficulties with Zen practice. His mind wandered uncontrollably. His legs gave him trouble, as one can imagine they would for a sedentary man his age. He also had a hard time understanding the rōshi's broken English. His perseverance shows poignantly in a photograph in Rick Fields's *How the Swans Came to the Lake*: an elderly figure in a Japanese robe over a white shirt with bow tie, sitting rather awkwardly in zazen.

In earlier letters to Mrs. Everett, Goddard urged her to meet Shigetsu Sasaki. This began a train of karma that has not slowed up a bit, for she became a key figure in the First Zen Institute that developed around Sasaki, a center that continues to be important in Western Zen Buddhism. Ultimately, Everett married Sasaki Osho shortly before he died. Her books, particularly *The Recorded Sayings of Lin-chi Hui-chao of Chen Prefecture*, edited from Sasaki's notes, and *Zen Dust: The History of the Koan and Koan Study in Lin-chi (Rinzai) Zen*, which she compiled with Isshū Miura, are essential references for serious Zen students.[14]

Though Goddard was one of five people to sign the original letter that requested that Shigetsu Sasaki be sent to the United States as a teacher, he was not convinced that Sasaki Osho's method of teaching was correct. With his experience in Chinese and Japanese monasteries, he felt that lay religious practice was vulnerable to worldly distraction and could not survive. He therefore endeavored to establish a monastic movement, the "Followers of Buddha." It was an ambitious project, set on forty acres in southern California adjacent to the Santa Barbara National Forest and also on a large parcel of rural land in Thetford. The religious brothers (no sisters) participating in the fellowship were to commute back and forth between the centers in a van, spending winters in California, summers in Vermont. The enterprise folded for lack of members. Rick Fields surmises—correctly, I think—that Goddard's strict monastic style went against the American grain, and his inability to persuade Wong Mou-lam or Wai-tao to head the movement left it without enlightened leadership.[15] It is ironic that despite his conviction that monasticism was the only possible path, his writings became an inspiration to Kerouac at the other end of the spectrum of lay and clerical practice and that his work fertilized the lay Zen Buddhist movement that flourishes today.

After immersing myself for several weeks in the project of piecing Goddard's life together, I find myself in the presence of a talented Yankee gentleman fired with bodhichitta, the aspiration for Buddhahood—who bewildered his conventional family and friends and worked a very lonely row quite single-mindedly. He knew that "Buddhahood" is not a sectarian matter, and one finds throughout his writings an aspiration to find the ultimate ground of religion—of whatever name.

The epigraph of his booklet *The Diamond Sūtra*, privately published in Thetford in 1935, is an adoration of the Three Bodies of the Buddha: the Dharmakāya (the pure and clear body), the Sambhogakāya (the full and complete body), and the Nirmānakāya (the infinitely varied body). The final section reads, "Adoration to Nirmanakaya: Buddhahood in its many bodies of manifestation—Shakyamuni Buddha, the perfectly Enlightened One; Jesus, the Nazarean and Saint Francis of Assisi; all the Bodhisattvas, Saints, and Sages of the past, present, and future; and Maitreya, the Coming Buddha."

This is not, as it might seem at first glance, a sentimental mixture of religions but a statement of devotion to the sacred perennial, which is manifested in many marvelous teachers. Goddard was a man of his time, the century that produced prophets of the perennial: Thoreau, Emerson, Whitman, Mary Baker Eddy, H. B. Blavatsky.[16] The World's Parliament of Religions at the Columbian Exposition in Chicago in 1893 was in some ways a fulfillment of their aspirations. John Henry Barrows, the leader of the parliament, wrote in "The History of the Parliament":

Religion, the white light of Heaven, has been broken into many-colored fragments by the prisms of men. One of the objects of the Parliament of Religions has been to change this many-colored radiance back into the white light of heavenly truth.[17]

Barrows goes on to quote lines from Tennyson about the many-colored fragments:

> Our little systems have their day;
>   They have their day and cease to be.
>   They are but broken lights of thee,
> And thou, O Lord, are more than they.

Many thinkers that late in the century could, however, skip over Tennyson, honor the many colors, and search for the pure light without giving it an immutable name and form.

Goddard sought the light in his own unique way. As Sasaki Oshō said, he distributed circulars. Writing was his Tao, his way of thinking through his changes, for himself and for others. His long task of compiling the biographies of *Eminent Engineers* surely gave him a sense of completion once his engineering career was over. He continued to write after he retired and during the years of working through the dark night of his earnest religious quest. In 1917, at the age of fifty-six, he published lectures he had given to students of the Chicago Theological Seminary under the title *The Divine Urge for Missionary Service*. Other works during this period include *The Good News of a Spiritual Realm* (a paraphrase of the Gospels), *Jesus and the Problem of Human Life*, and *Love in Creation and Redemption: A Study in the Teachings of Jesus Compared with Modern Thought*. He edited and published a journal, *The Good News of a Spiritual Life*, from 1918 to 1922.

After his exposure to Dr. Reichelt's views, Goddard explored the possibility that Buddhism might serve to inform Christianity. In 1924 he published the booklet *A Vision of Christian and Buddhist Fellowship in the Search for Light and Reality*, and in 1925, the eclectic metaphysical novel *A Nature Mystic's Clue*. These publications marked a development his former colleagues could not tolerate.

Alice Brannon noted in her chronology for 1924: "Rufus M. Jones forsakes him," and "Dr. Cadman of American Board disagrees." He was not swayed. He returned to Nanking for a second stay at Dr. Reichelt's monastery. In 1927, he published his inquiry: *Was Jesus Influenced by Buddhism? A Comparative Study of the Lives and Thoughts of Guatama and Jesus*.

Then, in 1928, Goddard made his pivotal trip to Japan where he studied with Suzuki and Yamazaki Rōshi. In 1930 Luzac of London published the book that emerged from these encounters, the first of his exclusively Buddhist works: *The Buddha's Golden Path: A Manual of Practical Buddhism Based on the Teachings and Practices of the Zen Sect, But Interpreted and Adapted to Meet Modern Conditions*. His journal, *Zen: A Magazine of Self-Realization*, later subtitled *A Buddhist Magazine*, appeared in a few issues, and the first edition of *A Buddhist Bible* was published in 1932. Then *Buddha, Truth, and Brotherhood: An Epitome of Many Buddhist Scriptures, Translated from the Japanese*, was published in 1934. During these years Goddard was also bringing out translations from Buddhist texts that were later incorporated in the second, enlarged edition of *A Buddhist Bible*.[18] The journal of Goddard's monastic movement, *Fellowship Following Buddha*, was published for a while. A total of twelve books and booklets appeared in the last fifteen years of his life. A few were commercially published; the rest were done at Goddard's expense, bearing prices of fifty cents or a dollar.

In hindsight one can trace the way his writing projects bore Goddard inexorably from applied science to Christianity to Buddhism and ultimately to Zen Buddhism. Though he settled on Zen as the prism that seemed least opaque, he remained true to his past. Kerouac heard that Goddard actually returned to Christianity in his final days.[19] In any case, as his outlook evolved, he became more inclusive. He did not fall into the error of isolating Zen from Buddhism or Buddhism from ethics and principles of kinship, and these principles were surely colored by his early Christian convictions. While *A Buddhist Bible* is true to the theme of the *Diamond Sūtra*— the Buddha cannot be known by any particular feature—at the same time, the various teachings of this collection clearly show that such a pure, profound understanding can be attained by the *practice* of Buddhism—the way of wisdom and compassion that Goddard sought to make his own. In a letter to Mrs. Everett he proclaimed, "[Zen Buddhism] is first, last, and always the practice of the Noble Path"[20]—that is, the Buddha's classical Eightfold Path of Right Views, Right Thoughts, Right Speech, Right Conduct, Right Livelihood, Right Effort or Lifestyle, Right Recollection, and Right Absorption or Concentration.

In 1937 and 1938 Goddard was deeply involved in matters relating to the publication of the second edition of *A Buddhist Bible*, which Alice Brannon noted was "his great contribution to the world's learning." He died on his seventy-eighth birthday in 1939.

In a memoir published in the June 1940 issue of *The Vermonter*, Charles R. Cummings wrote of Goddard's dislike of the telephone, his preference for plain food and secondhand clothing, his choice of a bus rather than a train for cross-country travel, and his opposition to hunting and fishing. He had a shrine room in his home, where he practiced zazen daily. "He developed a reputation as an eccentric, to say the least," one of my correspondents in Thetford wrote to me recently. As a fellow eccentric and fellow late-bloomer, I bow in veneration to the Sage of Vermont and to the life he devoted to religious understanding.

Foreword to *A Buddhist Bible*, by Dwight Goddard (Boston: Beacon Press, 1994).

THE CLASSICAL DISCOURSES

The Brahma Vihāras

T H E classical vihāra, or ancient Buddhist temple, was built of red sandalwood. It had thirty-two chambers and was eight tala trees in height. The tala tree is the palmyra palm, seventy or eighty feet tall, so the classical vihāra would have been as high as 640 feet. It was surrounded by beautiful landscaping with a bathing pool and had promenades for walking meditation. All creature comforts were provided in the furnishings, including stores of food, clothing, bedding, and medicine.

Though the classical vihāras were splendid temples, the Brahma Vihāras are far higher and richer. "Brahma" means "pure," and the Pure Vihāras can be understood as the Buddha realms of the noblest attitudes and conduct. The first of these Vihāras is *maitrī*, boundless kindliness; the second is *karunā*, boundless compassion; the third is *mudita*, boundless delight in the liberation of others; and the fourth is *upeksha*, boundless equanimity.[1]

Notice that "boundless" is the operant qualifier of these noble abodes. "Boundless" has come to mean "infinite," but its primary meaning is "without constraints." Boundless kindliness, compassion, goodwill, and equanimity are the noble qualities of the one who is no longer caged by individuality and has entered the spacious, multicentered universe.

Most of us, however, cannot readily find that open, inclusive space. We live without giving love easily, because openness and giving seem to endanger the precious self we cultivated as children. The

Buddha assures us, "We are all in this together. Let's trust one another, work through our fears, and build our Vihāras together."

It might seem to you that before you build your Brahma Vihāras, you should prepare your foundation—that is, sit hard, experience no-self—and then go on to apply your experience in daily life. But these are not ordinary temples. You prepare the ground at the same time that you build. Building is preparing; preparing is building. When you practice kindliness, you are also practicing no-self.

Kindliness is an attitude of pleasantness, interest in the other, encouragement. The kindly person is not worried about giving away personal power. In fact, when you are kindly, you are cultivating Buddha power, the power of decency that brings your interrelationship with all beings into clear focus.

A step beyond kindliness, compassion is the personal experience and practice of interbeing. We live our short lives not merely in interdependence but as a single great organism of many dynamic elements. What happens to you happens to me; what happens to me happens to you—at the same moment with the same intensity. If your behavior seems strange, it is because I am not yet well acquainted with your side of my psyche. I hear painful bondage in your angry words. I want to understand how it could have developed. I want you to hear my story too. Let's get together and share, and your part of me will become more clear—my part of you will become more clear.

Delight in the joy of others is still another step, more difficult to realize even than compassion. Each of us has a seed of personal potential, formed by a mysterious process of cause and affinity with ancestors and environment extending to unknown reaches of time and space. We visualize that seed maturing and bearing fruit, and when someone's attainment, however small, steals our thunder, we find ways to violate the Seventh Precept by praising ourselves and abusing the other.[2] Can the runner-up wholeheartedly congratulate the winner? That is the great test.

The final abode, equanimity, contains all the others, of course. Kindliness, compassion, and goodwill all rest and come forth here. Is it all right to be mortal? Is it all right to be homely? Is it all right to be weak in mathematics or grammar? Is it all right to be deaf to good music or blind to good art? Is it all right to be neurotic? Is it all right for others to have such egregious faults? Really all right—to the very bottom? Well, if so, then congratulations! "Golden-haired lion!" as Yün-men would exclaim. Please take my seat and be my teacher— be the Buddha's teacher.

I suspect, rather, that no one can answer in the affirmative and rightfully claim such broad, serene acceptance of self and others. The abode of equanimity stands vacant; nobody lives there, though some live right next door—the Buddha, Kuan-yin, the Dalai Lama (who stresses equanimity in his teaching and personifies it in his life). Yet in your practice the Buddha is sitting on your cushions and counting your breaths. Kuan-yin hears the sounds of suffering with no one's ears but your own and reaches out to help. These are not abstract figures but names and forms of yourself. In your own Brahma Vihāra, just as it is, you restore your equanimity and find energy to engage in the world with its beings.

The self is still present—but it is not self-preoccupied. It washes the dishes and puts them away. Even the ego is there, with a clear understanding of who this Buddha is and how she or he is engaged: crying the wail of others, laughing their laugh, and doing their work with them. You as your own Buddha can do this, quite short of any deep realization, *as if* you were doing it from your heart, for your heart is not just the little organ beating in your chest.

Thich Nhat Hanh has said that you are like a TV. If you want a peaceful channel, you can turn to a peaceful channel. If you want some other kind of channel, you can turn to that. Easier said than done, perhaps. My suggestion that you simply *be* compassionate and peaceful is also easier said than done sometimes. Perhaps you will need to search out conventional psychotherapy or to practice Theravada or Vajrayana exercises of self-assurance and loving-kindness. Treat these needs as you would your needs for the right food and plenty of exercise for good health.

A healthful regimen should not be regarded simply as a path to well-being, though one may become fit. The way of kindliness *as if* you were kindly is likewise not merely self-development, though the self does become more decent as you persevere. These skillful means are practice, but "practice" also is becoming a worn-out expression.

It is not merely practice to prepare oneself through study, zazen, and therapy to help others by suffering with them, delighting in them, and finding peace with them. At every step, at every point in your breath-by-breath return to your task, you are pursuing the Buddha's own noble, completely fulfilling work.

As the World-Honored One was walking with his followers, he pointed to the ground and said, "This spot is good for building a sanctuary."

Indra, the Emperor of the Gods, took a blade of grass, stuck it in the ground, and said, "The sanctuary is built."

The World-Honored One smiled.[3]

Simple for Indra, and thus simple for us. But just as Indra established a sanctuary in present place and time—just as the lotus flower blooms in the mud—so Brahma Vihāras rise in the midst of pain, anguish, affliction, and distress; of greed, hatred, and ignorance. Māra, the destroyer, usually considered to be the Buddha's opposite number, dwells there in the four ignoble abodes of suspicion, antipathy, jealousy, and restlessness. Māra is not, of course, merely an outside influence. He is really quite ordinary looking, tall, getting a little soft around the tummy, with rather unkempt hair and a white goatee. However, I vow to cut off the Three Poisons of greed, hatred, and ignorance; that is, I vow to follow my Buddha tendencies of kindliness, compassion, goodwill, and peace, rather than to follow my Māra tendencies.

It is, as Indra showed, quite easy. It is also quite difficult. You must cut off the mind road, as Wu-men says. Otherwise, "You are like a ghost, clinging to bushes and grasses."[4] Yet you and I know from experience that the mind road can't be cut off by even the strongest, sharpest power of the will. In fact, the more energy you devote to confronting your mental chatter, the noisier it becomes.

Don't misdirect yourself. Attend scrupulously to your task, and respond as appropriately as you can to its demands. Though your thoughts are importunate, on your cushions your task is the kōan Mu, and its requirements are devotion and the utmost spirit of inquiry. When you doubt that you are giving Mu enough devotion and inquiry, then indeed you are not. Settle into Mu and forget about your inadequacies. If the Buddha had been adequate in his zazen, he would not have had to practice so long. Inadequacy, like the Three Poisons, is your field of noble endeavor. Without inadequacy, there is no Buddha Way.

In daily life, your noble endeavor is likewise attention and response. Listening to the sounds of the world, you find that your very skin is a sensory organ rather than some kind of outer bulwark. "Inside and outside become one, and you are like a mute person who has had a dream."[5] It is then that the thrush sings and the earth is shaken. You find yourself an organ of interbeing, and turning the wheel of the Dharma becomes a matter of waving hello to your neighbor when you pick up the morning paper. Your concern about an illness in your neighbor's family is in that wave. Your pleasure at her purchase of a cabin in the country is in that wave. Your peace of mind is in that wave.

This is the great peace of equanimity. Once you see into the vast and fathomless emptiness of the universe, you find there is no mind road to cut off and no abiding self to protect. The Three Poisons are wiped away, and although they can appear again, at least you have learned how to turn the switch. You know clearly for yourself the royal ease of Kuan-yin.

Kuan-yin must nonetheless be served with food, clothing, bedding, medicine, and love. Otherwise he or she cannot lift a hand to wave. Māra too demands service; otherwise he or she cannot be suspicious, antagonistic, jealous, and restless. The Buddha, Kuan-yin, and Māra are useful archetypes, but I am the one who must take responsibility. If I meet most of my needs and allow others to help me to meet them, then I can be of service. If I meet most of my self-centered desires and allow others to help me to meet them, then I will get bloated and everybody else had better watch out.

Ordinary law is at work here. If you follow this law as Buddha, then it is clearly the Buddha's law. If you follow it as Māra, it is still the Buddha's law but not so evidently, for it is misused. The law is, of course, karma. *This* happens because *that* happens. *This* is because *that* is.

The misuse of karma is not its denial. The most hardened criminal knows that one thing leads to another. Misuse lies in the attempt to manipulate and exploit karma for self-centered purposes. When parents are too preoccupied to care for their children, then the children tend to neglect themselves and their responsibilities in turn. Schools are disrupted, hospitals are filled with victims of accidents and addictions, and prisons are crowded with violators of progressively more severe laws.

On the world stage, when agricultural syndicates buy up the land of peasants for sugarcane or bananas, then the restless peasants will have to be neutralized somehow, and the insects and diseases that come with specializing in a single crop will have to be dealt with. The problems that inevitably come with neutralizing the peasants and with dealing with insects and plant diseases will have to be addressed in turn, using ever more draconian measures.

The synergy of draconian misuse becomes overwhelming and can only be dealt with by the radical transformation of the individual self from an isolated being to a multicentered being and by leadership from that position, which speaks to the needs of the family, community, and the universal organism. Gary Snyder wrote an article more than twenty years ago entitled "Buddhism and the Coming Revolution."[6] It was based on the Hua-yen model of the Net of Indra and showed clearly how the Mahayana Buddhist, faithful to the intimate interdependence of all beings, resists the ordinary self-aggrandizing tendencies of states, institutions, and individuals. It was a seminal essay, and we are challenged to bring its thought into reality in our practice.

This practice is all of a piece. The tendencies of some governments to torture political dissidents and destroy the rain forests are my own tendencies and yours—to neglect our children, say, or to squander our physical health. The results are felt all of a piece too. My Māra permeates the world, and world Māra permeates me. Likewise when I conduct myself with kindliness, compassion, goodwill, and equanimity, all beings are enlightened.

The path is practical, not metaphysical, though archetypes show through the mist to guide us. I cannot be content just to seek peace of mind with concepts. After all, the Buddha's teachings were concerned with noble conduct, and the ancient sanghas were organized as noble social structures. How should we as modern Buddhists apply the teaching and sangha models in the context of pervasive systems that are ignoble and destructive? It is up to us, it seems to me, to conspire with our families and friends and establish practical alternatives that are true to our understanding, here in our own time.

1990

Emmei Jikku Kannon Gyō

The Ten-Verse Kannon Sūtra of Timeless Life

Kanzeon	Kanzeon!
namu butsu	I bow before the Buddha,
yo butsu u in	with the Buddha I have my source,
yo butsu u en	with the Buddha I have affinity—
buppō sō en	affinity with Buddha, Dharma, Sangha,
jō raku ga jō	constancy, ease, assurance, purity.
chō nen kanzeon	Mornings my thought is Kanzeon,
bo nen kanzeon	evenings my thought is Kanzeon,
nen nen jū shin ki	rapidly thoughts arise in the mind,
nen nen fu ri shin.	thought after thought is not separate from mind.

In the Diamond Sangha we recite the *Emmei Jikku Kannon Gyō* in Sino-Japanese,[1] though we tried out an English translation in the early days of the Maui Zendō. That experiment lasted two days as I recall and was firmly put down.

Yamada Kōun Rōshi once remarked that the *Emmei Jikku Kannon Gyō* is really a kind of dhāranī, or invocation of praise. I think he was right, offering a clue, perhaps, to the reason my English translation didn't work. Far Eastern dhāranīs are not translations but transliterations of Sanskrit. Original texts are lost for the most part.[2]

54

Their rhythmic forms have been recited for 1,500 years, and we sense how they have been empowered with the devotions of thousands of monks, nuns, and laypeople over the centuries.

The *Emmei Jikku Kannon Gyō* was, however, composed in Chinese, not in Sanskrit. It has the rhythm and the empowerment of a dhāranī, but its rational meaning becomes evident.

To begin at the beginning: the original title was simply *Jikku Kannon Gyō*, "The Ten-Verse Kannon [Kuan-yin] Sūtra."[3] *Jikku* (or *jukku*) means "ten verses" or "ten lines." *Gyō* (or *kyō*) is "sūtra." The term *emmei* (technically *enmei*—but the *n* is elided to *m*) was added to the title sometime after the sūtra was composed and is usually interpreted to mean "prolonging life." Zen teachers including Hakuin Ekaku Zenji are fond of recounting folk stories about the miraculous effects of the sūtra as a lifesaver.[4]

It is important, however, to see into other implications of *emmei* and such related terms as *kotobuki* (long life) and *furui* (ancient). The ultimately ancient is the timeless, a reference to the essential nature that is not born and does not die. It is timeless—not eternal and not ephemeral. The Morning Star presented the timeless to the Buddha Shākyamuni, and distant peach blossoms presented the timeless to Ling-yün.[5] The *Emmei Jikku Kannon Gyō* extends life for the believer, and it brings the Buddha's own experience of timelessness in this moment to the rest of us.

The first line evokes Kanzeon (Kuan-shih-yin), the full name of Kannon, "The One Who Perceives the Sounds of the World." Kanzeon evolved from the male figure Avalokiteshvara in the earlier Indian Buddhist pantheon and is worshiped as the "Goddess of Mercy" in Far Eastern folk religion.

As the Goddess of Mercy, Kanzeon saves people from drowning, injury, injustice, evil spirits, and bandits. She releases prisoners whether they are innocent or guilty, and gives pregnant women their choice of a male or a female child. If you fall off a cliff and call out Kanzeon's name, you will be suspended in the air like the sun—or so we are assured by the *Lotus Sūtra*.[6]

I honor this deep faith. It is a comforting bulwark that sustains countless people who might otherwise be miserable and hopeless.

You will even find Kanzeon worshipers among Zen students. But Harada Dai'un Rōshi would ask, "How old is Kanzeon?" If the student replied, "Ageless," the rōshi would ring his little bell. "Back to your cushions for more work!" Kanzeon may be a goddess in the sky who manipulates karma for the true believer, but *she* can also be *he*— looking out through very ordinary eyes, hearing with very ordinary ears, using very ordinary tools like a wok or a word processor.

Indeed, sometimes Kanzeon is represented holding a thousand tools in a thousand hands. Sometimes she stands with a compassionate expression, holding a jar of ambrosia and a lotus. Sometimes she is just sitting there, altogether comfortable in her mudra of royal ease, with one arm over her upraised knee. Her presence alone serves as a teaching. With many mudras, she teaches in many ways—for example: "Time for supper!"

When Mother or Father calls from the kitchen, "Time for supper!" the children cry out, "Time for supper!" This echo is not just Kanzeon's announcement bouncing about but the children's confirmation that they have internalized the message and have made it their own.

Wu-men, hearing the sound of the drum announcing the noon meal, found himself to be the Peak of Wonder, dancing at the center of paradise.[7] He then announced himself in a great career of teaching that set generations of subsequent teachers dancing. This is the function of Kanzeon.

"Ripeness is all," so of course the function of Kanzeon is also to study and practice—to prepare for the drum or the Morning Star. The drum and the Morning Star also had to prepare for their announcements. This is not a pathetic fallacy. The timpanist plays upon a living being. The stars are bursting with their messages. Turn to a child for the star's announcement.

Enlightening, being enlightened, calling and responding, the birds and stars as Kanzeon save us, just as they themselves save us and as Kanzeon the figure "up there in the sky" saves us. At the same time, it is important to heed Yün-men's cautions:

> Yün-men addressed his assembly and said, "To realize the Way through hearing voices; to enlighten the mind through seeing

colors—that is as if the Bodhisattva Kuan-shih-yin [Kan-zeon] bought cheap cakes with her penny. If she throws them away, she gets *man-t'ou* instead."[8]

Man-t'ou (Japanese: *manjū*) is elegant steamed pastry. Yün-men is reminding us that we mustn't cling to the realization we find on hearing or observing something, as profound as it might be. Realization in the course of Zen practice is important, but it is just a preliminary experience. Stopping there is as though Kanzeon were to deny herself her fulfillment.

It is also important not to cling to *man-t'ou*—that is, to the more complete experiences that come later—or even to the archetype "Kanzeon" herself. Individuation on the Buddha Way is the triumph of selfless practice—and the unspoken, unacknowledged triumph of forgetting triumph, and forgetting that you have forgotten. It is an endless process of realizing more and more fully the wisdom of Yün-men's caution, "It is better to have nothing than to have something good."[9]

Kanzeon is not the one who perceives the sounds of the world, as the *Diamond Sūtra* might say if it addressed the matter—therefore we call her "Kanzeon." Therefore we begin the *Emmei Jikku Kannon Gyō* free of concepts by calling her name: *"Kanzeon!"*

This is an evocation, it seems to me, like old Western poets calling upon their muse to inspire their lyrics. We call upon Kanzeon to inspire our sūtra and our lives. We call upon ourselves as Kanzeon. We call upon ourselves to inspire Kanzeon.

With this evocation, the sūtra sets forth the foundation of Kanzeon's being: *namu butsu*. *Butsu* is Buddha—"the Enlightened One," Shākyamuni, and each of his wise and compassionate successors—a name empowered even more than the sūtra itself with the devotions of countless disciples. The Buddha is also the wisdom and compassion that comes forth as I and you. It is the name we give to enlightenment and its potential and to the mystery that brings forth beings and their universe.

Namu is transliterated from the Sanskrit *namah* and means literally "to submit to," "to make obeisance to." With our understanding of obeisance we can translate it, "I cast everything away in the presence of the Buddha." As Dōgen Zenji says, the ocean of essential nature is the most important matter. "It is beyond explanation. We can only accept it with respect and gratitude."[10]

Acceptance is Kanzeon, hearing the sounds of the world. She perceives the distress and pain everywhere and is realized by the announcements of geckos and children. The compassionate action of Kanzeon arises from the empty place of grateful receiving. With *namu butsu* I venerate the great power for the Way that is generated by this profound act of opening myself: "I bow before the Buddha."

Yo butsu u in / yo butsu u en. These two lines present my relationship with the Buddha, and yours—our relationship with this great power for the Way. An incomplete translation of the two lines would be "With the Buddha we have *in*; / with the Buddha we have *en*." *In* and *en* are specifically Buddhist terms that mean "direct or inner cause" and "indirect or environmental cause."

Every action occurs in the harmony of *in* and *en*, and the two terms are often combined to reflect this unity as *innen*. The tree grows from its seed—the direct cause—and with the influences of earth, water, air, and sun—the indirect causes.

Like Kanzeon, as Kanzeon—you and I have *in*, direct cause, with the Buddha. We form the cause of Buddha. "This very body is the Buddha."[11] This transitory, imperfect frame is the Tathāgata, the Buddha who comes forth *thus*, no less than Shākyamuni himself.

How to translate *in* is the question. I choose *source*, the best I can do, but it is a tautological rendering. It is like saying the body is the source of the body. The Buddha is the source of the Buddha. But sometimes tautology can be instructive as an agent of emphasis, to bring the matter home.

You and I also have *en*, indirect cause, with the Buddha. *En* is a very important word in the Japanese language. *Fushigi na en*, "mysterious affinity," is an intimate expression frequently used to confirm a relationship between teacher and student or between friends or lovers. "Affinity" is a word we can use to describe how molecules come together to form a cell, how cells come together to form the organs of a fetus. The dynamics of the plenum are the dance of affinity and separation. It is affinity with the Buddha that brings the affinity of the Sangha. Sangha affinity realizes the Buddha.

Buppō sō en—affinity with Buddha, Dharma, Sangha, the Three Treasures. Affinity with the Buddha is repeated here—it cannot be repeated too often.

Dharma has many meanings, more than forty, I am told, but for our purposes there are two: "teachings" and "practice."

The teachings have three aspects:

1. The universe and its beings have no identifiable substance, either in form or in essence.
2. Each being is in intimate communication with all other beings, for indeed each is a constituent of all others.
3. Each being is unique and stands forth alone in its containment of the infinite variety in the plenum.

The practice too has three main elements:

1. Following the guidance of a good teacher and the writings and examples of good teachers of the past.
2. Following the guidance of friends, family members, and all beings.
3. Following the guidance of personal inspiration.

The teaching without sound practice is abstract and ultimately meaningless. Practice that is not based on sound teaching can be self-centered and capricious. Bracing the union of teaching and practice is the clear sense of responsibility that we inherit directly from the Buddha, and bracing this sense of responsibility is bodhichitta, the imperative for realization and compassion. Without this organic chain of inner directives, the Dharma is neglected and society and its habitat slide into corruption.

Finally, we have affinity with the Sangha. *Sangha* comes from the Sanskrit root meaning "aggregate" and has been used traditionally to refer to the Buddhist priesthood. In the Mahayana, however, it refers to all beings and their symbiosis—their interbeing. In the sūtra, Sangha is the "Buddha Sangha," the kinship of students who practice together, the only mode of realizing the Buddha Way. There are Buddhist hermits, of course, but these are men and women who choose solitude after long years of practicing with sisters and brothers. They are firmly grounded in their realization of intimacy and containment. Note also that Sangha is the harmony of the Buddha and the Dharma, the practice and realization of evanescence, intimacy, and diversity in my life and yours, and in our lives together.

We have our source in and our affinity with the Buddha, we find affinity with the Three Treasures of Buddha, Dharma, and Sangha, and—like Kanzeon, as Kanzeon—with this affinity we have *jō raku ga jō*, "constancy, ease, assurance, purity." These are the four attributes of nirvana—freedom from time, distress, bondage, and delusion.[12]

Nirvana is freedom, but not anything antinomian. Freedom is play. By "play," I mean enjoyment, of course, but also scope for movement—flexibility. You and I are not irrevocably linked to a relentless sequence of events, and therefore we can find timelessness in the moment. We do not feel doomed, and therefore "every day is a good day."[13] We must feed, clothe, shelter, and medicate ourselves, but we can also devote ourselves selflessly to family, community, and the world. We can enjoy the world without exploiting it, and we need not isolate ourselves. I am fond of the line from the *Ts'ai-ken-t'an*: "Water which is too pure has no fish."[14]

Jō, the first of the four attributes of nirvana, is a translation of a Sanskrit term meaning "constant" or "eternal," but it also means "normal" or "ordinary." The timeless is the ordinary. The changeless is the everyday.

> Chao-chou asked Nan-ch'üan: "What is the Tao?"
> Nan-ch'üan said, "Ordinary mind is the Tao."

After further conversation, Nan-ch'üan goes on to say, "If you have truly realized the Tao beyond doubt, you will find it is as vast and boundless as outer space."[15] Just like the sky! How ordinary! How everyday!

Vast as outer space—the Tao is not created and not extinguished. It is the unmoving void, charged with inconceivably dynamic potential, bursting through with Chao-chou's realization—or standing up when guests enter the room, making greetings in the most cordial and hospitable manner. *Jō* is Kanzeon, like smoke that is neither empty nor substantial—timeless essential nature passing around the cookies and juice.

Raku, the second attribute, is comfort or ease, as in Kanzeon's mudra of royal ease. I am always quoting Nakagawa Sōen Rōshi: "Enjoy your Mu." This seems difficult when legs hurt and anxiety comes up, but the rōshi is reminding us that practice is enlightenment—the nature of training is the nature of realization. At the beginning, zazen is commonly stressful, but it mellows into something quite comfortable before realization appears. Practice before and after that experience is Kanzeon herself, comfortably sitting there, with mind and body at rest. This is not, of course, to encourage indulgence. Have a look at Hakuin's portrait of Kanzeon sitting comfortably in zazen. She is not neglectful.

I translate *ga*, the third attribute, as "assurance," but literally the ideograph means "self." As the Dalai Lama is always saying, assurance and equanimity are the same thing. Equanimity is the quality of mind that comes with realizing to the very bottom that everything is empty and that emptiness is truly all right.

In terms of karma and science, selves are made up of the Five Skandhas, perceptions and things perceived, which in turn arise from genes and early influences that are ultimately untraceable. Metaphysics once again leads to a dead end, perhaps leaving one ready for understanding that the skandhas are themselves empty and that the genes and environmental influences are reducible to formless, ultimately unidentifiable tendencies of the void.

Nonetheless, here I am! Here is the Buddha! Here are the many beings! The equitable self is the presentation of the equitable universe, potent with confidence as the plenum itself is confident.

We are creatures of the great forest of the cosmos. We are born, grow through childhood and youth into maturity, and pass through old age and death, leaving our remains to compost the vitality that continues to pulse and burgeon. My role and yours are one: to produce the richest compost we can, not only for the future but as an ongoing product of the many small deaths of our lives.

Small deaths, the Great Death of realization, and physical death —these are all a matter of relinquishment, of forgetting the self. It is the self forgotten that is equitable, assured, realized.

Tao-hsin made his bows before Seng-ts'an and said, "I beg the compassion of Your Reverence. Please teach me the Dharma way of emancipation."

Seng-ts'an said, "Who is binding you?"

Tao-hsin said, "No one is binding me."

Seng-ts'an said, "Then why should you search for emancipation?" Hearing this, Tao-hsin had great realization.[16]

The delusion that he was bound was the bondage of Tao-hsin. Freedom from that bondage was Tao-hsin realized. It sounds simple, but better men and women than any of us have sweat blood to make this great transition from the self center to the multicenter, from the *me* world to the compassionate world of Kanzeon.

Jō, the final attribute of nirvana, is different from *jō*, the first word. (There are scores of words pronounced *jō* in Japanese, all of them with different graphs.) This *jō* means "purity," including the ordinary sense of "clean."

Some people are preoccupied with purity, and this is their defilement. In the old days of the New Age, I met many of them, dying of fruitarianism, breaking out in boils they regarded as drains for their impurities. Thank heaven we are past that phase!

Hui-k'o, Bodhidharma's successor, helped Seng-ts'an through his preoccupation with purity:

Seng-ts'an said, "Your disciple is suffering from a fatal illness [probably leprosy]. I beg Your Reverence to release me from my sins."

Hui-k'o said, "Bring me your sins and I will release them for you."

Seng-ts'an said, "I cannot find my sins, in spite of all my efforts."

Hui-k'o said, "I have already released your sins for you. From now on, live by the Buddha, Dharma, and Sangha."[17]

With this, Seng-ts'an found liberation, perhaps not from his physical symptoms but from his tenacious preoccupations, and in turn he was able to help Tao-hsin to find freedom from his—in the remarkably similar exchange quoted earlier. Leprosy was, and still is, widely believed to be the result of bad karma, and Seng-ts'an shared this impure idea. Actually, however, karma is not some kind of judge out there waiting to condemn us. When Seng-ts'an looked closely, he couldn't find his bad karma. Hui-k'o confirmed his freedom from such delusions and cautioned him to practice his realization in daily life by maintaining the Buddha Way.

All this is not to deny the effects of one's conduct. If Seng-ts'an had lived in poverty in a district where leprosy was prevalent, then he was probably more vulnerable to the disease than other people might have been. It would, however, be a crude oversimplification to suppose that he contracted the disease because, say, he threw a stone at a leper when he was a little boy in an earlier incarnation.

Of course, an act of throwing a stone at a leper becomes part of the subtly complex network of affinities relating to leprosy that pervades the universe. But to draw a direct causal connection from throwing the stone to catching the disease is like other deterministic ideas about karma—for example, that a little white thing escapes from the body at death and goes cruising around looking for a couple making love where it can find its rebirth. Such notions reduce to crude materialism the untraceable, infinitely dynamic streams that come together to form the infant, and so we say they are superstitions.

At the same time, Seng-ts'an and you and I are responsible for our cruel conduct. When people get hurt, our atonement will not mend broken bones, but it can transmute anger, revenge, and self-condemnation. At the beginning of our sūtra service in the Diamond Sangha, we recite:

> All the evil karma ever created by me since of old,
> on account of my beginningless greed, hatred, and ignorance,
> born of my conduct, speech, and thought,
> I now confess, openly and fully.[18]

Chō nen kanzeon / bo nen kanzeon. *Chō* is morning and *bo* is evening. "Mornings my thought is Kanzeon, evenings my thought is Kanzeon." (English is such a discursive language! See how succinctly the lines read in Sino-Japanese: "Morning thought Kanzeon, evening thought Kanzeon.") Of course, "thought" doesn't mean "musing on." If the line were rendered, "Mornings I think of Kanzeon," that would place Kanzeon back on the altar. It is rather, "Mornings my thought is no other than Kanzeon herself or himself."

The graph for *nen*, "thought," is made up of two parts: the top is "present" or "now"; the bottom is "mind." Together they are the vast mind coming forth in this moment, framed by this particular personality and situation.

The "vast mind" is a discovery of the Mahayana and does not appear in Classical Buddhism, where "mind," however subtle and powerful, is the human mind and nothing more.[19] The Mahayana vision is a paradigm shift, a discernment of mind as void of attributes or fixed nature yet the source and nature of everything, "whether as sentient beings or as Buddhas, as rivers and mountains of the world"[20]—and, of course as *nen*, as thoughts.

Coming forth, the *nen* is a pulse or a frame, and the topic is supplied by past circumstances, often immediately past. Here the sūtra says that every moment, morning and night, my thought is itself the Bodhisattva enlightened, enlightening the self and enlightening all beings.

Like our vows, the *Emmei Jikku Kannon Gyō* sets forth a light for our path. My best person is Kanzeon—each moment of perceiving a bulbul, of listening to complaints about my failings, of speaking out against injustice and cruelty, of breathing Mu on my cushions.

Nen nen jū shin ki / nen nen fu ri shin. "Rapidly thoughts arise in the mind; / thought after thought is not separate from mind." This final couplet sets forth the human talent of thinking and its function as an expression of mind.

There are two dictionary meanings for *nen nen*. The first is "one sixty-fifth of a second." Once I experienced this rapidity of thought as I sat in zazen. I found myself facing a large, very soft, organic-looking mass that was pulsing with life. There was a little aperture in this mass like a mouth or a vagina, and tiny beings were escaping from the aperture one after another in very rapid succession. I knew these tiny beings were my thoughts, and somehow this vision helped me to deepen my practice.

In the first line of the final couplet of the sūtra, the second character for *nen* is not simply the first one repeated but a graph that functions like our ditto mark. Thus the two words would be literally translated "thought-thought," implying "thought-thought-thought-thought . . ." ad infinitum. This is how thoughts arise in the mind.

The second dictionary meaning of *nen nen* is "thought after thought." Here the focus is on the particular thoughts as they appear, rather than upon their rapidity. Thus in the second line of the couplet the repeat sign is not used, but the graph for *nen* is itself repeated. The implication is that each thought as it arises is the mind itself, and in this instance, conveying Kanzeon herself or himself. So I translate *nen nen* in the first line to indicate their rapidity in emerging in the individual mind, and in the second line to reflect their nature as particular expressions of something infinitely larger.

These two lines of the final couplet can be traced to *The Awakening of Faith in the Mahayana*, a succinct summation of doctrine that probably originated in China in the sixth century. In Yoshito S. Hakeda's translation we read:

> Though it is said that there is [an inception of] the rising of [deluded] thoughts in the mind, there is no inception as such that can be known [as being independent of Mind].[21]

The interpolations are those of the translator, and in the setting of the text, I would judge them to be accurate. The notion of a universal yet empty mind arose with the Mahayana and by the sixth century it could be formulated quite clearly in *The Awakening of Faith*.[22] Thoughts were earlier considered to be irrevocably delusive, but now they could be seen as being the mind itself. Still later this formulation was internalized by Zen Buddhist monks. Bassui Tokushō, a fourteenth-century Japanese Zen master, addressed his assembly and said:

> The mind is intrinsically pure. When we are born it is not created, and when we die it does not perish. It has no distinction of male or female, nor has it coloration of any kind. . . . Yet countless thoughts issue from this . . . nature as waves arise in the ocean.[23]

If thoughts arise like waves in the ocean, they also follow the same path as light, in particles of waves—in waves of particles. Thoughts are the photons of the mind. They rise and fall endlessly, but if we focus only upon the flow, the continuum becomes simply a time line, ending finally for us when interventions in the hospital fail. Too bad.

In the experience set forth by this sūtra, each thought, ever so brief, is Kanzeon herself, turning the Dharma wheel. I am reminded of the image of Kuya Shōnin at Rokuharamitsuji in Kyoto. Kuya was an early Pure Land teacher who went about the country calling out the mantra of his faith, *"Namu amida butsu,"* "Veneration to Amitābha Buddha." The sculptor renders him as a young pilgrim walking slightly bent over, with a row of little Amitābha Buddhas coming out of his mouth on a wire.

All those little Buddhas coming from Kuya's mouth look exactly the same, but the many Kanzeon thoughts that emerge as the mind are different. Some are full of praise, some are critical; some are jolly, some are sad. Some say, "Let me help you." Some say, "You must do it yourself." Moreover, as William James makes clear, each thought qualifies one's previous thought and also qualifies one's environment and personality and worldview as well as the environment as it is changing with the thoughts expressed by others.[24]

In my vision of thoughts appearing from a little aperture, I could identify the large, living mass as my own mind. I have had other visions I could identify as my mind, particularly as a structure of cells greatly enlarged and distantly removed like very high clouds, as though I were viewing them from the inside, as the transparent ceiling of the world. These visions help to verify the message of the sūtra—the mind is the source of thoughts, and the brain of the individual human being is not the only reference. George Meredith said that stars are the brain of heaven.[25] Would people and birds be the brain of Earth? To respond, one must begin with fundamentals:

P'an-shan said, "In the Three Worlds there is no dharma. Where shall we search for the mind?"[26]

Dharma, with a small *d*, is phenomena. I understand P'an-shan to be suggesting that in the Three Worlds of past, present, future and all dimensions, there is nothing at all to be called mind or anything else. The old teachers said the same kind of thing about one's personal mind. In *The Zen Forest* we read, "The nature of my mind is the empty sky."[27] Then where do thoughts come from? Where do people and birds come from?

Frankly, I don't know, but beneath the airy ceiling of my great skull I find many thoughts that sing of speckled eggs and ropes and things, not to mention Robert Louis Stevenson. The tiny beings I saw emerging from the unknown aperture of my living flesh re-present trees, towers, animals, and people of a landscape that changes from Australia to Hawai'i to California, though it is always intimately familiar.

Yamada Kōun Rōshi used to say, "The mind is empty infinity, infinite emptiness, full of possibilities." For human beings these possibilities can be ignorant or imbued with song, but it is the singing *nen* that changes the world. Children understand this in their own way very well and resonate to Stevenson's "Singing":

> Of speckled eggs the birdie sings
> And nests among the trees;
> The sailor sings of ropes and things
> In ships upon the sea.
>
> The children sing in far Japan,
> The children sing in Spain;
> The organ with the organ man
> Is singing in the rain.[28]

Hui-neng says that with an enlightened thought, you compensate for a thousand years of evil and destruction.[29] This is the potency of what we call provisionally the Great Mind, brought forth in a single pulse. Wu-men says the one-moment *nen* perceives eternity,[30] but the *Emmei Jikku Kannon Gyō* finally does away with perception. One tiny thought is itself the mind of the mountains, the rivers, the great Earth, the sun, the moon, and the stars[31]—the mind of all beings.

Realizing this is *com*-passion—living, suffering, and dealing with exigencies together with everyone and everything. It is Kanzeon's realization, her light on my path, and though there is no way

I can consistently be the Kanzeon as drawn by Hakuin or carved by Mokujiki or set forth in the *Emmei Jikku Kannon Gyō*, I can find her as myself in peak moments, as you can, and otherwise we can be our own best Kanzeon, open to songs and flowers, showing peace in a troubled world.

Kanzeon! Namu Butsu!

1989

The Virtue of Abuse

Tōrei Zenji's "Bodhisattva's Vow"

TRANSLATION AND COMMENTARY

I *am only a simple disciple, but I offer these respectful words*:

When I regard the true nature of the many dharmas,[1] I find them all to be sacred forms of the Tathāgata's never-failing essence. Each particle of matter, each moment, is indeed no other than the Tathāgata's inexpressible radiance.

With this realization, and with compassionate minds and hearts, our virtuous ancestors gave tender care to beasts and birds. Among us, in our own daily lives, who is not reverently grateful for the protections of life: food, drink, and clothing! Though they are inanimate things, they are nonetheless the warm flesh and blood, the merciful incarnations of Buddha.

All the more, we can be especially sympathetic and affectionate with foolish people, particularly with someone who becomes a sworn enemy and persecutes us with abusive language. That very abuse conveys the Buddha's boundless loving-kindness. It is a compassionate device to liberate us entirely from the mean-spirited delusions we have built up with our wrongful conduct from the beginningless past.

With our open response to such abuse we completely relinquish ourselves, and the most profound and pure faith arises. At the peak of each thought a lotus flower opens, and on each

flower there is revealed a Buddha. Everywhere is the Pure Land in its beauty. We see fully the Tathāgata's radiant light right where we stand.

May we retain this mind and extend it throughout the world so that we and all beings become mature in Buddha's wisdom.[2]

Tōrei Zenji is Tōrei Enji (1621–92), disciple of Hakuin Ekaku and de facto founder of Ryūtaku Monastery in Mishima, Japan.[3] His *Bodhisattva's Vow* is part of the traditional Rinzai sūtra collection,[4] but among the various Rinzai temples it is included in daily services only at those that are associated with him, so far as I know. Both Katsuki Sekida, our first Diamond Sangha teacher, and I practiced at Ryūtakuji, and thus it was natural for us to include the piece in our own services, using Mr. Sekida's translation. Recently I composed this new version, with the help of Yamaguchi Ryōzō Oshō, priest of the Myōshinji branch temple on Maui.[5]

The theme of Tōrei's homily is the sacred nature of all phenomena, the particular effectiveness of abusive words in bringing one to a realization of the Buddha Dharma, and a vow that all beings share this realization. One can distinguish five phases of his theme:

1. The deceptive modesty of Tōrei's introductory words is reminiscent of the complete absence of ego found in the utterances of so many teachers in Zen literature. They are not merely expressions of self-effacement but of complete vacancy: the Dharmakāya, vast emptiness. There is nothing at all from the very beginning. Purity of the mind is "not a single thing." When all forms are abandoned, there is the Buddha, but even the Buddha can in no way be identified. If the mind depends on anything, it has no sure haven.[6]

2. "Each particle of matter, each moment, is indeed no other than the Tathāgata's inexpressible radiance." It is from the brilliant void, *as* the brilliant void, Tōrei Zenji says, that phenomena appear as the sacred forms of the Tathāgata's essence. "Tathāgata" is a synonym for "Buddha," but it is an altogether neutral term, meaning simply, "Thus come," or "The one who thus comes." As the *one*, it is at rest, but when it appears as each form, each particle of matter, each sense experience, in each moment—it reveals itself, where we stand, as inexpressibly radiant light. Ikkyū Zenji wrote:

Striking bamboo one morning he forgot all he knew.
Hearing the bell at fifth watch, his many doubts vanished.
The ancients all became Buddhas right where they stood.
T'ao Yüan-ming alone just knit his brows.

Ikkyū is citing stories that were well known in his time about old teachers who found that all voices are the voices of the Buddha. Also, he is suggesting that readiness is all, and some, even the great poet T'ao Yüan-ming, aren't ready.[7]

Be clear about this. Fo-yen Ch'ing-yüan, a Sung period master, remarked:

Ancient Zen teachers were so compassionate that they said, "Activity is Buddha activity, sitting is Buddha sitting, all things are Buddha teachings, all sounds are Buddha voices." It is, however, a misunderstanding to think that this means all sounds are actually the voice of enlightenment, or that all forms are actually forms of enlightenment.[8]

This means, I think, that all sounds are actually the voices of the Buddha, but only when we ourselves are actual. I cite again the well-known story of T'ou-tzu and his importunate monk:

A monk said to T'ou-tzu, "'All sounds are the sounds of Buddha'—is that correct or not?"

T'ou-tzu said, "Correct."

The monk said, "Doesn't your asshole make farting sounds?" T'ou-tzu hit him.

The monk said, "'Coarse words and gentle phrases all have their source in essential nature'—is that correct or not?"

T'ou-tzu said, "Correct."

The monk said, "Then may I call Your Reverence a donkey?" T'ou-tzu hit him.[9]

Yüan-wu says, "Too bad for this monk, he had a head but no tail."[10] In other words, he had intellectual understanding but no visceral realization. If you presume to denounce someone as a donkey, you had better be confident of your salvific power.

3. "With this realization, and with compassionate minds and hearts, our virtuous ancestors gave tender care to beasts and birds."

Back through the ages to the Buddha Shākyamuni himself, the old teachers were receptive to the many dharmas, and they moved sensitively in nature, nurturing animals and plants, seeing, hearing, feeling the Tathāgata everywhere.[11] And sensing the possibilities of such experiences, surely all of us are profoundly grateful for our food, drink, clothing, and medicine, as the warm flesh and blood of the Buddha. Tōrei Zenji's text uses the term *raihai gasshō*—deep bows, hands held reverently palm to palm, to describe the thankful spirit that we maintain. Pots and pans are Buddha's body.

4. "All the more, we can be especially sympathetic and affectionate with foolish people, particularly with someone who becomes a sworn enemy and persecutes us with abusive language." This is not merely Tōrei Zenji's challenge but a perennial summons—to be so wise and compassionate as to "suffer fools gladly," as Paul urged. This can be bitter medicine. Bergen Evans, a profound student of Western culture through its popular sayings, remarks that today we not only don't accept Paul's injunction but quite the contrary, consider it a mark of wisdom *not* to suffer fools gladly at all.[12]

Sad to say, Evans is correct; we have strayed from the ancient. Yet Tōrei Zenji does not merely urge us to tolerate fools but points out incisively that their "very abuse conveys the Buddha's boundless loving-kindness. It is a compassionate device to liberate us entirely from the mean-spirited delusions we have built up with our wrongful conduct from the beginningless past." This carries us into the profound depths of the *Diamond Sūtra*:

> If you are scornfully reviled by others, this misery is the result
> of your harmful acts in previous ages. But with this scorn and
> vilification from others, your evil karma from previous ages
> is extinguished.[13]

Again and again, from earliest times, we find this rigorous guidance repeated in our tradition. In Buddhist metaphysics, *k'ung-jen* is the patience that is based on the realization that everything is empty.[14] But not only are the Buddha's trials insubstantial, they are joyous. In the *Parinirvāna Brief Admonitions Sūtra*, the Buddha cautions his monks to practice the ultimate patience that transforms the poison of abuse to the nectar of the gods.[15]

Early in the eighth century Yung-chia wrote:

Let the abusive words of others pass;
they try to set fire to the heavens, and will only exhaust
    themselves.
I hear their abuse as though I were drinking ambrosia;
everything melts, and I enter the place beyond thought and
    words.[16]

Yung-chia echoes the Buddha's theme of abuse as ambrosia, and
Musō Soseki echoes Yung-chia's metaphor of melting with the
abuse:

> People's abuse
>   has melted what was golden
>   and it has gone from the world
> Fortune and misfortune
>   both belong to the land
>   of dreams[17]

Even that which is golden melts away, and Musō enters the place
beyond fortune and misfortune. Thus it is the abuse itself, as a sacred
particular, that clearly and incisively comes forth as the very point of
the Tathāgata—and like Tung-shan Shou-ch'u, you slap your knees
and cry out, "That's it!"

Yün-men said to Tung-shan, "Oh, you rice bag! Do you go
about in such a way, now west of the river, now south of the
lake!" Tung-shan was deeply enlightened.[18]

In one of his talks to lay followers, the Japanese master Bankei
Yōtaku, an exact contemporary of Tōrei Zenji, tells an interesting
story of response to abuse that brings fine talk down to everyday
practicality:

I have a temple in Edo, located in Azabu on the outskirts of the
city. We once had a man there who worked around the temple.
He had an interest in religious matters to begin with, I think,
for he was always observing the daily lives of the monks. From
this, a genuine religious aspiration must have developed in
him naturally. In any case, one evening some of the monks
sent him on an errand that took him to the outer fringes of the
city where houses were few and far between. It was an area

where from time to time a samurai wanting to try the edge of his blade on a human body had been appearing and cutting down passing travelers. The monks were concerned for his safety because it was getting dark and he would have to pass through this dangerous area. But he told them not to worry, and he set right out, saying that he would be back soon. As the messenger returned in the growing darkness, however, sure enough the samurai stepped out at his usual haunt and brushed past him.

"You brushed your sleeve against me on purpose," he growled, drawing his sword.

"But my sleeve didn't even touch you," replied the messenger. Then, for some reason, he prostrated himself before the samurai three times. The samurai, who had raised his sword and was on the point of striking, now unaccountably lowered it.

"You're a strange one," he said. "Well, go on, I'll let you pass." And the messenger escaped unharmed.

Now a tradesman had seen all this take place. He had fled to the safety of a nearby roadside teahouse and had witnessed the events from his place of hiding.

When he saw the sword about to fall, he turned his eyes away and waited fearfully for the inevitable to happen. When he finally looked up again, he saw to his surprise that the messenger was standing right before him.

"You certainly got out of that by the skin of your teeth!" he said. He then asked the messenger what had made him think to give the three bows.

The messenger answered that all the people where he worked bowed three times. "My mind was completely empty. I just thought, if you're going to strike me with that sword, then do it. I made those bows without thinking. The man told me I was a strange fellow and said he would spare me. Then he allowed me to go past."

So, having barely escaped death, the messenger returned safely to the temple. I told him that I thought this was because of the depth of his religious mind, which enabled him to reach the heart of such a lawless samurai. It goes to show that nothing is more trustworthy than the Buddhist Dharma.[19]

Indeed—although as Yang-shan remarks in effect to Kuei-shan, even the Buddha Dharma itself is untrustworthy, and there is nothing whatever to rely upon.[20] The messenger's mind was truly deep, and he had relinquished himself with his humble response. There with his pure faith a lotus flower opened, and the Buddha appeared. The Pure Land was displayed in the village that evening.

The messenger's particular response to the threat of violence fitted the circumstances like the lid to the box, but it was for a particular box and wouldn't necessarily fit another one. Sometimes expedient resistance to violence can reveal the way of the Buddha. Sometimes neither resistance nor relinquishment will prevent a dreadful crime. I am sure, for example, that twelve years ago, when three Maryknoll nuns and a lay missioner were faced with rape and death, they drew passionately on all their religious resources. But they were not spared. The Pure Land by whatever name was nonetheless displayed in the countryside and in every country, as it was once on Calvary. Perhaps after the crime they reached the hearts of their assassins.[21]

5. "May we retain this mind and extend it throughout the world, so that we and all beings can become mature in Buddha's wisdom." Here the Bodhisattva reaches the place of total trust in nothing at all and finds there the glorious light of the Pure Land, and vows that this light may shine brilliantly in the conscious mind of all beings. May the light of Tōrei Zenji's mind illumine our own minds—across the sea and across the centuries. May his great power in turning the wheel of the Dharma give spirited energy to our own resolve.

1992

PRACTICE

The Way of Dōgen Zenji

H ᴇᴇ -Jin Kim's *Dōgen Kigen: Mystical Realist* was the first com-
prehensive study in English of Dōgen Zenji's writings, and for the
past twelve years it has served as the principal English-language reference for those Dōgen scholars who work from his thirteenth-
century Japanese and for Western Zen students reading translations
of his writings. This new edition appears in a scholarly setting that
now includes many new translations and studies of Dōgen, and thus
it is most welcome.

   Dōgen wrote at the outermost edge of human communication,
touching with every sentence such mysteries as self and other, self
and no-self, meditation and realization, the temporal and the timeless, forms and the void. He moved freely from the acceptance of a
particular mode as complete in itself to an acknowledgment of its
complementarity with others, to a presentation of its unity with all
things—and back again. He wrote of the attitude necessary for understanding, of the practice required, of the various insights that
emerge, and of the many pitfalls. He did not generally write for beginners—most of his points require very careful study, and a few of
them elude almost everybody. These challenges are compounded by
his creative use of the Japanese language of his time. It has been said
that he wrote in "Dōgenese," for he made verbs of nouns, nouns of
verbs, created new metaphors, and manipulated old sayings to present his particular understanding.

   Thus the writings of Dōgen are an immense challenge to anyone seeking to explicate them in English, but Kim does a masterful job. I do not presume to explicate Kim's words but to offer a personal perspective of Dōgen in hopes that it might serve as access to Kim's incisive scholarship.

I choose as my theme a key passage in the "Genjō Kōan," the essay that Dōgen placed at the head of his great collection of talks and essays, the *Shōbōgenzō*, and I use Kim's translation here:

> To study the Way is to study the self. To study the self is to forget the self. To forget the self is to be enlightened by all things of the universe. To be enlightened by all things of the universe is to cast off the body and mind of the self as well as those of others. Even the traces of enlightenment are wiped out, and life with traceless enlightenment is continued forever and ever.[1]

To study the Way is to study the self. Asian languages offer the same options as English for the meaning of the verb "study." To paraphrase dictionary definitions, it is "to examine with the intention of learning." Thus I would interpret Dōgen's words: "To come to understand the Way is to come to understand the self."

The term "Way" is a translation of *Dō* in Japanese, *Tao* in Chinese. It is the ideograph used to identify the central doctrine of Taoism and its basic text, the *Tao-te ching*. Kumarajiva and his colleagues in the early fifth century selected *Tao* as a translation of Dharma, a key Sanskrit Buddhist term meaning "law," or "way of the universe and its phenomena," or simply "phenomena." In Dōgen's view, all phenomena are the Buddha Dharma—the way of the universe as understood through Buddhist practice.

Indeed, for Dōgen, to study and understand the Buddha Way is to practice the Buddha Way, and to practice the Buddha Way is to have the self practice. It is important to understand that practice is both action and attainment. Modes of practice—zazen, realization, and the careful work beyond realization—all these are complete in themselves, and they are also means for further completion. They are acts of particular moments, and they are also stages in the course of time.

As to the self, it has no abiding nature and, in Blake's words, "kisses the Joy as it flies." It is the Buddha coming forth now as a woman, now as a youth, now as a child, now as an old man, now as an animal, a plant, or a cloud. However, animals and plants and clouds cannot "study" in Dōgen's sense, so in this context, Dōgen refers to the human being that can focus the self and make personal the vast and fathomless void, the infinitely varied beings, and their marvelous harmony.

To study the self is to forget the self. Here Dōgen sets forth the nature of practice. My teacher, Yamada Kōun Rōshi, has said, "Zen practice is a matter of forgetting the self in the act of uniting with something." To unite with something is to find it altogether vivid—like the thrush, say, singing in the guava grove. There is just that song, a point of no dimension—of cosmic dimension. The "sole self" is forgotten. This is something like the athlete who is completely involved in catching the ball, freed of self-doubt and thoughts of attainment, at the same time aware of the other players and their positions. Using this same human ability on one's meditation cushions is the great Way of realization. It must be distinguished from thinking *about* something. When you are occupied in thinking, you are shrouded by your thoughts, and the universe is shut out.

There are other analogies for gathering oneself in a single act of religious practice, freeing oneself of doubt and attainment. Simone Weil sets forth the academic analogy: contemplating an object fixedly with the mind, asking myself "What is it?" without thinking of any other object relating to it or to anything else, for hours on end.[2]

Dōgen often uses the phrase "mustering the body and mind" to understand oneself and the world. Using Kim's translation of a later passage in the "Genjō Kōan":

> Mustering our bodies and minds we see things, and mustering our bodies and minds we hear sounds, thereby we understand them intimately. However, it is not like a reflection dwelling in the mirror, nor is it like the moon and the water. As one side is illumined, the other is darkened.[3]

This mustering is zazen—and also the activity of the Zen student who is grounded in zazen. Kim quotes Dōgen writing elsewhere in the *Shōbōgenzō*:

The Buddhas and Tathāgatas have an ancient way—unequaled and natural—to transmit the wondrous Dharma through personal encounter and to realize supreme enlightenment. As it is imparted impeccably from Buddha to Buddha, its criterion is the samādhi of self-fulfilling activity.

For playing joyfully in such a samādhi, the upright sitting in meditation is the right gate.[4]

With the practice of zazen, mustering body and mind, we understand a thing intimately by seeing or hearing, and the self is forgotten. This kind of understanding is not by simile, it is not a representation, like the moon reflected in the water, but is a brilliant presentation of the thing itself and is a complete personal acceptance. One side is illumined. There is only that thrush. At the same time, the universe is present in the shadow. The other players are still there.

To forget the self is to be enlightened by all things of the universe. The term "enlightened" is *shō*, the same *shō* found in *inka shōmei*, the document given to a senior student by a master confirming him or her as a teacher. The thrush confirms you, enlightens you, but be careful not to give "enlightenment" anything more than provisional status. It is likely to be just a peep into the nature of things. Nonetheless, "one impulse from a vernal wood" or the Morning Star shining over the Bodhi tree is a communication. The communication works the other way, from the self to the object, but the result is different, as Dōgen makes clear earlier in the "Genjō Kōan":

> That the self advances and confirms the myriad things is called delusion; that the myriad things advance and confirm the self is enlightenment.[5]

The way of research and analysis is "called" delusion. Don't condemn it, Dōgen is saying. By advancing and confirming and throwing light on all things of the universe, you reach intellectual understanding. However, when you forget yourself in mustering body and mind in the act of practice, there is only that particular act, in that particular breath-moment. Then, as Kim says, the whole universe is created in and through that act. With this you experience the things of the universe. They are your confirmation, your enlightenment.

To be enlightened by all things of the universe is to cast off the body and mind of the self as well as those of others. Focusing body and mind with all one's inquiring spirit on a single matter, the self is forgotten. The myriad things communicate their wisdom with their forms and sounds, and the emptiness, harmony, and uniqueness of the ephemeral self and the world are understood clearly. This is reminiscent of Paul's "putting off the old man"—not merely forgetting but dying to the self.

Casting off body and mind should not be confused with self-denial. Many people suppose that they must get rid of the self. The Buddha too went through a phase of asceticism, avoiding food and sleep in an effort to overcome his desires. Such a path has a dead end, as the Buddha and others have found. We need food and sleep in order to cast off body and mind. The Way is gnostic rather than ascetic.

Finally, as Dōgen says, when you cast off body and mind, all other beings have the same experience. One version of the Buddha's exclamation under the Bodhi tree reads, "I and all beings have at this moment entered the Way!" This does not mean "All beings can now come along." Rather, at the Buddha's experience, all beings simultaneously cast off body and mind.

> When Hsüeh-feng and Yen-t'ou were on pilgrimage together, they became snowbound in the village of Wushantien. This gave them time for an extended dialogue, during which Hsüeh-feng recounted his various spiritual experiences. Yen-t'ou exclaimed, "Haven't you heard the old saying, 'What enters from the gate [that is, by intellection] cannot be the family treasure?'" Hsüeh-feng suddenly had deep realization and exclaimed, "At this moment, Wushantien has become enlightened!"[6]

With his exclamation, Yen-t'ou cast off body and mind. Simultaneously, Hsüeh-feng did the same. Personalizing Bell's theorem a thousand years and more before Bell, the whole village was likewise affected.

Even the traces of enlightenment are wiped out, and life with traceless enlightenment is continued forever and ever. Wiping away the intimations of pride that come with a realization experience are the ultimate steps of Zen practice, steps that never end. They form the Way of the Bodhisattva, polishing the mind of compassion, engaging in the travail of the world, "entering the marketplace with bliss-bestowing hands." Over and over in kōan practice, the Zen student works through the lesson of casting off, casting off.

A monk said to Chao-chou, "I have just entered this monas-tery. Please teach me."
Chao-chou said, "Have you eaten your rice gruel?"
The monk said, "Yes, I have."
Chao-chou said, "Wash your bowl."[7]

"Have you eaten your essential food?" "Yes, I have." "If so, wipe that idea of attainment away!" For our present limited purposes this would be an explication of Chao-chou's meaning. What is left after body and mind are cast off? Endlessly casting off—ongoing prac-tice. The "Genjō Kōan" ends with the story:

When the Zen teacher Pao-che of Ma-ku was fanning him-self, a monk asked him, "The nature of wind is constant, and there is no place it does not reach. Why then do you fan yourself?"
Pao-che said, "You only know that the nature of wind is constant. You don't yet know the meaning of its reaching every place."
The monk asked, "What is the meaning of its reaching every place?"
Pao-che only fanned himself. The monk bowed deeply.

The nature of the wind is Buddha Nature, "pervading the whole uni-verse." The monk's question is an old one. If all beings by nature are Buddha, why should one strive for enlightenment? Dōgen himself asked such a question in his youth, and his doubts fueled his search for a true teacher. Pao-che takes the monk's words, "reaching every place," as a figure of speech for Zen Buddhist practice that brings forth what is already there. As Dōgen says in his comment to this story—the final words of the "Genjō Kōan":

Confirmation of the Buddha Dharma, the correct transmis-sion of the vital Way, is like this. If you say that one should not use a fan because the wind is constant, that there will be a wind even when one does not use a fan, then you fail to understand either constancy or the nature of the wind. It is because the nature of the wind is constant that the wind of the Buddha House brings forth the gold of the earth and ripens the kefir of the long river.

The wind of the Buddha House—the practice of zazen, realization, and going beyond realization—is altogether in accord with the wind of the universe, the Buddha Mind. As Dōgen says elsewhere, "The Dharma wheel turns from the beginning. There is neither surplus nor lack. The whole universe is moistened with nectar, and the truth is ready to harvest."[8] The harvesting of truth, the practice of forgetting the self, the practice of realizing forms and sounds intimately, the practice of polishing our mind of compassion—this is our joyous task.

Foreword to *Dōgen Kigen: Mystical Realist*, by Hee-Jin Kim (Tucson: University of Arizona Press, 1987).

Ultimate Reality and
the Experience of Nirvana

Satori and Shūnyatā

As a teacher of Zen Buddhism, I confess that I feel a little uneasy with my assigned topic. I find such terms as "ultimate reality" and "nirvana" to be abstract and unreal—absolutes that fit philosophical schemes, perhaps, but not the requirements of Zen students who face the challenge of maturing as human beings in their practice.

"Satori," now an English word, thanks to its introduction by D. T. Suzuki, has come to imply omniscient wisdom. I much prefer the term *kenshō* (Chinese: *chien-sheng*), which holds the more moderate meaning of "seeing into (essential) nature." Shūnyatā, the void, expresses deepest experience, but I find that all too readily it becomes something abstract called "nothing."

I spend time with inquirers disabusing them about absolutes. When someone who has read a little in Zen Buddhism asks me if I am enlightened, I respond without hesitation that I most certainly am not. When someone asks me how many kōans I have passed, I respond that I am still working on my very first kōan and haven't passed it yet. This is not false modesty but is true to the very bottom. There is enlightenment beyond enlightenment, passing beyond passing. Each milestone on the path may seem a be-all and end-all experience. Everything falls away. The everyday self disappears. Yet the path continues to open out.

Experience is the moment; the path is endless practice. They are like the frame and the narrative of a movie. The student glimpses the timeless in the frame, but the movie continues. Frame and movie are like the complementary principles of light: without the photon, there is no wave of light. Without the frame, there is no movie.

"Movie" can be a helpful metaphor, but it is limited. The complementarity realized in Zen Buddhist practice is not confined just to time with no time but includes form with emptiness, the mundane with the spiritual, the particular with the universal, and the dimension of birth and death with the dimension of no-birth and no-death.

The Buddha Shākyamuni taught this apparently complex yet actually very simple complementarity more than 2,500 years ago. With the passage of his teaching through many cultures and languages, the original manner and expression of his Way have evolved significantly in a variety of directions. Yet the archetypal message is the same: "Human beings tend to be miserable because they are preoccupied with themselves. When they are free of their self-centeredness they can find happiness."

That is to say, you and I tend to get absorbed in patterns. We tend to become fixated on the temporal, the mundane, the particular, and the world of being born and dying. When we see into the nature of things and make intimate the formless, the timeless, the spiritual, the universal, the world of no-birth and no-death, then we are evolving on the path to full and complete lives. We can see for ourselves how our previous views were correct, yet only so far as they went. We once saw the world of forms and time but not their essential qualities of no-form and no-time. We gave meaning to the many people, animals, plants, and things in our richly varied world, but now those many beings give meaning to us. Thus we can be liberated from constrictions that bind us to an atomized existence.

This liberation brings happiness—not simply self-contained happiness but the joy of work in the world. It is not mere adjustment to the needs of others but an extinction of the acquisitive self, a peak experience that must then be processed. This new practice is not merely a kind of cognitive realignment but a dynamic engagement in consequential possibilities.

The Buddha's basic teaching is usually and rightfully summed up with reference to the Four Noble Truths, traditionally considered to be the content of his first sermon: (1) anguish is everywhere, (2) there is a source of anguish, (3) anguish can be rooted out completely, and (4) the path of this liberation is eightfold: Right Views, Right Thinking, Right Speech, Right Action, Right Livelihood, Right Effort or Lifestyle, Right Recollection, and Right Meditation. The term "right" should be understood to mean "in accord with the essentially vacant, interdependent, and richly varied nature of things."

Anguish, the Buddha said, has its root in clinging to the notion of a permanent and independent self or soul. When it becomes clear that the self and indeed all things are not only evanescent but illusory and that everyone and everything come into being interdependently, then one is liberated from the misery that comes with a preoccupation with "me and mine."

The entire corpus of Buddhism can be seen as practice and realization of this simple formula of the Four Noble Truths. My purpose here is to show how Zen Buddhism is one of its many particular developments, how it is being applied today, and what applications might be appropriate in the future.

To begin with, the Pali term *dukkha* (Sanskrit, *duhkha*), rendered here as "anguish," is not well served by its usual translation: "suffering." I agree with Walpola Rahula that "suffering" is an ambiguous English word that is not an appropriate translation, for it can mean "enduring" and "allowing" as well as "experiencing pain." The word "anguish," it seems to me, sums up the Buddha's allusion to the profound sense of dissatisfaction felt by human beings about their dependence on others and about the transitory nature of their lives and of everything around them, particularly their possessions and structures.[1]

Liberation from this profound dissatisfaction does not, of course, come by waving a wand. The Eightfold Path is a rigorous way to liberation, with a scrupulous and exacting formation of views, speech, conduct, and practice. It is the perfection of character, with "perfection" understood as a process rather than a state. The realizations along the way are profound and transformative, but the end is not yet.

Zen Buddhists have mined the Four Noble Truths and the Eightfold Path for treasure, and the outcome is twofold: (1) a particular practice of meditation that leads to Right Views of the world and its beings as evanescent and essentially harmonious and (2) a daily-life practice that brings essential harmony into worldly reality. Compassionate modes of attitude, speech, and conduct lead to Right Meditation, and Right Meditation leads to further compassionate modes of attitude, speech, and conduct.

In the meditation hall, the Zen Buddhist student is encouraged to muster body and mind and focus on single points, one at a time. The preliminary practice, by no means an easy one, is to count the breaths. The breath is the spirit, as traditional peoples across the world understand. While Buddhists will not isolate the spirit as an entity, the metaphor is nonetheless useful. With counting the breaths, one links body, brain, spirit, and will.

The exercise is to count inhalations and exhalations both, or just the exhalations, from "one" to "ten." If we return to the metaphor of a movie, then each number in this counting is the individual frame in the epic of counting breaths. The end of all epics is expiration, once and for all. Thus, in Zen Buddhist practice, and indeed in any religious practice worthy of the name, one's attention is not particularly devoted to sequence. The object is not to reach "ten" so much as it is to become intimate with each point as it comes up—with just that point "one," just that point "two," just that point "three," in the whole world. Everything else in the mind is quiet. "Intimacy" and "realization" are synonyms in traditional Zen Buddhist texts. The point has no dimension, no magnitude. *There* is the timeless itself. *There* is the universal; *there* is the dimension of no-birth and no-death.

With some sense of the possibilities of meditation, the student can move on to other exercises. In the Sōtō school of Zen Buddhism, usually this will be the rigorous practice of facing the timeless void.

In the Rinzai school, the student will be given cases from the literature to face in meditation—not to analyze but to confront and make intimate. In both options the way is one of understanding, of taking upon oneself. There is no abstraction here, no philosophy of religion.

Take, for example again, the seminal story of Chao-chou's dog:

A monk asked Chao-chou, "Has the dog Buddha Nature or not?"

Chao-chou said, "*Mu.*"[2]

Probably Chao-chou said, "Mu." It is thought that the word was pronounced in such a way in his time. In any case, it is modern Japanese pronunciation and has passed in that way into use at North American and European Zen Buddhist centers. The word means "No," or "Does not have."

Clear enough, but if "does not have" were the sole meaning of Chao-chou's response, the entire practice of Mahayana Buddhism would be thrown into confusion, for the literature plainly states that all beings have, or indeed *are*, Buddha Nature. The monk really is asking, "What is Buddha Nature?" So if Chao-chou is not denying Buddha Nature, then he is either temporizing or he is somehow affirming it.

It is clear from the many commentaries on this case that Chao-chou is not merely temporizing. He is saying in effect, "You are really asking what Buddha Nature is. Well, I'll tell you: *Mu.*"[3]

So the question is, then, "What is Mu?" This is the point to which the student musters body and mind. In his comment on Mu, Wu-men urges his students to carry the word day and night, concentrating on it with their "360 bones and joints and their 84,000 hair follicles," with all their inquiring spirit. With earnest, one-pointed practice, Wu-men promises, you will find your own ground, and even the Buddha and his great successors had better stay out of your way.[4]

There is a risk, however, of getting stuck in Buddha Nature— that is, in the void that is simply potent with all things. The well-known story of the hundred-foot pole is a cautionary tale in this respect:

> Ch'ang-sha had a monk ask Master Hui, "How was it before you met Nan-ch'üan?"
> Hui just sat there silently.[5]

Ch'ang-sha and Hui were once brother monks in Nan-ch'üan's assembly. Ch'ang-sha became the teacher at a large monastery, while Hui secluded himself in a mountain hut. Ch'ang-sha wondered how his old friend was getting along, so he sent a monk to see him, after priming the monk with a leading question. Hui responded to the question by not responding. The dialogue continued:

> The monk asked, "How was it after you met Nan-ch'üan?"
> Hui said, "There couldn't be anything different."

In other words, before his experience with his teacher, Hui found himself in empty silence, and after his experience, he still found himself in empty silence. The monk returned and told Ch'ang-sha about this conversation. Ch'ang-sha came forth with a poem by way of comment:

> You who sit on the top of a hundred-foot pole,
> although you have entered the way, it is not yet genuine.
> Take a step from the top of the pole
> and worlds of the ten directions will be your entire body.

The top of the hundred-foot pole is the isolation of Hui in a selfless condition. He has experienced one side of the complementarity of form and emptiness, but he has not integrated the two aspects of reality for himself, as himself. Even after meeting the great Nan-ch'üan, he is still stuck in the void.

"Take a step from the top of the pole." This is the test point of the case, which students through the centuries since Ch'ang-sha have presented to their teachers. For our purposes, we can see how Ch'ang-sha is emphasizing the importance of moving on from simple awareness of the insubstantial nature of the self and all things. With that step, "worlds of the ten directions will be your entire body." That is, you will find mountains, rivers, the great Earth itself, the sun, the moon, the stars, people, animals, plants, streets, and towers to be your own great self. The monk then challenges Ch'ang-sha:

"How can I step from the top of a hundred-foot pole?"
Ch'ang-sha said, "Mountains of Lang, rivers of Li."
The monk said, "I don't understand."
Ch'ang-sha said, "The four seas and the five lakes are all under imperial rule."

"The mountains of Lang, the rivers of Li, the four seas and five lakes" are specifics of worlds of the ten directions. Who is the emperor here? At another time, Ch'ang-sha enlarged on his principal point:

> The entire universe is your eye; the entire universe is your complete body; the entire universe is your own luminance. The entire universe is within your own luminance. In the entire universe there is no one who is not your own self.[6]

Not only is there no *one* who is not myself or yourself, there is nothing at all that is not each of us. No leaf, no stone, no gecko that is not I myself, you yourself. Thus the self arises—not merely interdependently with all things but *as* all things. It is all things—interbeing, to use Thich Nhat Hanh's expression.

The photon and the wave theories are just preliminary and conflicting insights into the reality of light. Likewise form and emptiness, however profoundly experienced, can be seen as steps to realizing the interbeing that gives them relevance. Interbeing, the uniqueness of form, and the void are the three-part complementarity that has been and continues to be explored to the depths by Zen teachers:

> A monk asked Ta-lung, "The body of form perishes. What is the eternal body?"[7]

Perhaps the monk is thinking that the eternal body is something absolutely empty or something absolutely solid. Like Ch'ang-sha, Ta-lung responded with a verse:

> The mountain flowers bloom like brocade;
> the river between the hills runs blue as indigo.

When it is clear that the absolute is none other than this lovely, rich world in its many forms—when the world and its animals and plants and people are found to be one's own body—then we walk with everybody and everything on a common path. This is compassion, suffering with others. "Suffering" is here an appropriate word. We endure, we allow pain and sorrow, we welcome gray hair, weakened powers, and death itself with our friends and family members.

Compassion is thus the liberation from self-preoccupation. The joy of this release and the simultaneous experience of inclusion bring forth a vow to work with the world with one's own hands—not imposing from above, not missionizing to redeem nonbelievers, but like Gandhi, weaving cloth for clothing with the village women. Or like Whitman, sitting with the wounded in a Washington hospital.

Walking a common path we realize more and more intimately how closely dependent we are on all people, animals, and plants, and how closely dependent they are on us. Like Hui's experience of emptiness, however, this cannot be simply a static disclosure—we cannot remain stuck there in a blancmange of oneness. Engagement the noun is engagement the verb—the practice. This is the Way of the Bodhisattva, who vows to postpone full and complete enlightenment for herself until all beings are enlightened. Turning the wheel of the Dharma, she feels in her bones and marrow the sounds of agony that William Blake heard and expressed so vividly in his poem "London":

> How the Chimney-sweeper's cry
> Every black'ning Church appalls;
> And the hapless Soldier's sigh
> Runs in blood down Palace walls.

> But most thro' midnight streets I hear
> How the youthful Harlot's curse
> Blasts the new-born Infant's tear,
> And blights with plagues the Marriage hearse.

The Bodhisattva Kuan-yin, who by her very name discerns the sounds of the world, is here the poet, experiencing inner-city cries as outer-city torment, then taking the most appropriate action—to

write a timeless poem about human anguish and, by clear implication, human responsibility. Such Bodhisattvas are rare, East and
West.

In Asia, cultural influences have in the past generally confined the
Bodhisattva ideal and imperative to doctrine and to the sangha and
the surrounding lay community. Exceptions can be readily cited,
from the Buddhist King Ashoka, who set forth decrees of human civility once he had carved out his kingdom, to the monk Gyōgi Bosatsu, who traveled around medieval Japan building waterworks for
the peasants. While Buddhists in general, including Zen Buddhists,
could from very early times rightfully be honored for their study of
many constructs of morality and for their practice of morality in ritual, meditation, and interaction with lay supporters, they could also
be criticized for avoiding social analysis and any application of their
understanding much beyond their temples and paths of pilgrimage.

Today such criticism is still valid to a certain degree. However,
many of the monastery walls are down, and where they are intact,
they have become quite porous. The superstition that Buddhists do
not get involved in politics is likewise disappearing, and across the
Buddhist world we find broad applications of the Buddha's teaching.
Some North American Zen Buddhist centers sponsor programs of
peace, justice, social and medical care, community organization,
bioregional organization, and the protection of nature. Participants
in these programs find inner guidance from their own experience of
dynamic unity with all beings and inspiration from such outstand-
ing thinkers and leaders as Joanna Macy and Gary Snyder. The
members also look for leadership to geniuses of engaged Buddhism
in South and Southeast Asia, whose names may not be familiar to the
average American or European.

Buddhadāsa Bhikkhu, for example, the late Siamese Buddhist
master, has challenged monks, nuns, and lay followers to restructure
society to be in keeping with natural balance and fundamental Buddhist teaching. Other key Southern Buddhist leaders who apply
their religious understanding in the world include Maha Goshananda, the leader of Cambodian Buddhists, who leads peace walks
in his own country along roads made dangerous with land mines, as
well as beyond his country, to show the way of peace to the world;

Aung San Suu Kyi, the elected president of Myanmar and a Buddhist, who has remained steadfast in her insistence on democracy in the face of arrest and imprisonment; A. T. Ariyaratne, who founded and inspires the Sarvodaya Shramadana movement of village self-help in Sri Lanka; Sulak Sivaraksa, who founded and perseveres in the face of government prosecution to guide the International Network of Engaged Buddhists and a number of other progressive associations in Siam; Thich Nhat Hanh, who coined the term "engaged Buddhism" and was a prominent figure in the peace process during the war in Vietnam, and who continues to maintain effective support for sufferers in his homeland. The one Asian Mahayana figure who serves as a wonderful model of engaged Buddhism for Zen Buddhists and Buddhists generally is the Dalai Lama, who lectures on loving-kindness without using a single Buddhist buzzword and who resolutely supports movements for human rights and the protection of nature—a prophet for everyone, regardless of their religion.[8]

It should be noted that the engaged Buddhism that is advanced by most of these North American and Asian figures is not mere service, though certainly service is an important element of their work, from the protection of refugees to the rescue of prostitutes. The teachings and writings of Buddhadāsa Bhikkhu on "Dhammic Socialism" have inspired the development of cooperative Buddhist communities in Siam, which Tavivat Puntarigvivat relates to the Base Communities of liberation theology in Latin America.[9] Studying Dr. Puntarigvivat's account of these Siamese Buddhist communities, I am struck by their reformative nature and at the same time their conservatism, for like the villages of the Sarvodaya Shramadana movement in Sri Lanka, they flourish from the root of traditional teachings.

The Buddhist Peace Fellowship (BPF) in North America, whose leadership is largely (though by no means completely) Zen Buddhist, has established a program called the Buddhist Alliance for Social Engagement, whose acronym (BASE) is clearly an echo of liberation theology and whose policy arises from the Buddhist teaching of dāna, or giving. Young volunteers live together with a schedule of sharing and religious practice while serving as apprentice workers in social welfare and medical agencies. At the same time, however,

Buddhist Peace Fellowship members are beginning to examine the futility of an engaged Buddhism that is limited to hospice and other medical and social welfare work. Speakers at a recent BPF institute challenged the leadership to consider how they might be functioning as no more than Band-Aids to the acquisitive system and that they might even be perpetuating its evils by helping it to work better.

The world is in a terrible mess. Great self-perpetuating economic and governmental institutions, fueled by the Three Poisons of greed, hatred, and ignorance, are contaminating vast populations of people, animals, and plants. Almost a hundred years ago, William Butler Yeats in his "Second Coming" asked the foreboding question:

> And what rough beast, its hour come round at last,
> Slouches towards Bethlehem to be born?

We already know the answer. The monster has been born, and we read of his foul depredations across the world in our daily papers. As a final point in this discussion of Zen Buddhist experience and its application, I suggest that Zen Buddhists, Buddhists generally, and men and women of all religions—and those of simple goodwill with perhaps no formal religion—face the task of finding a way, perhaps like the folks in self-reliant villages and ashrams of South and Southeast Asia, to live in this society but not of it and to network like Buddhist sanghas of classical times to create a new way of life that is at the same time as old as the world, a way that is grounded in gracious generosity.

Revised from a paper read at the colloquium: "Buddhism and Christianity: Convergence and Divergence." Pontifical Council for Ecumenical and Interreligious Affairs, Fo Kuang Shan, Kaohsiung, Taiwan, R.O.C., July 31–August 4, 1995.

Ritual and Makyō

I HAVE not emphasized ritual and ceremony in my teaching very much. I suppose the reason for this lies partly in my own humanist and Protestant heritage, partly in the nature of my training under the monk Nyogen Senzaki, Yasutani Haku'un Rōshi, and Yamada Kōun Rōshi, and partly in my interpretation of Western culture and my assumptions about what might be suitable for Western students.

In my childhood and youth in Honolulu, I attended Sunday school and adult services at Central Union Church, the Congregational church that descendants of New England missionaries established in early days. I was fairly content about attending because I met my friends there, and I didn't imagine any other options. My response to the classes and services was about the same as my response to school at that time—boredom—and again, I didn't imagine that any other response might be possible. My mother was secretary to a succession of ministers of the church, and my father was a member of the standing committee—equivalent to a board of directors. Neither was particularly religious, and I can recall my father remarking that the chief function of religion was to maintain the social fabric of the community. I didn't know any other religion, and probably "our way" of avoiding anything ritualistic was established in my mind in opposition to "their way" of kneeling and reciting "Hail Mary," which I heard about vaguely from my classmates in school.

My first encounter with Buddhism was R. H. Blyth's *Zen in English Literature and Oriental Classics*, and Dr. Blyth was my first teacher, there in the internment camp we shared. He tended to make fun of ritual, and I sensed that he put up with it in order to get what he considered to be the essential teaching.

After the war, I met the monk Nyogen Senzaki, who came to the United States in 1905 and worked in a variety of jobs before establishing himself as a teacher of Zen Buddhism in 1925. During those twenty years he learned a lot about American culture and was convinced that ritual was not very suitable for American Zen students. Accordingly, our only ceremony besides the basic arrangement for doing zazen together was reciting the "Four Vows" in Sino-Japanese at the end of the meeting and then taking tea together.

Later I visited Japan again and studied with Nakagawa Sōen Rōshi, and after Senzaki Sensei's death, Anne Aitken and I both became Sōen Rōshi's disciples. With regard to ritual, his path was completely different from that of Senzaki, for his entire life was ritualized, from his act of looking at the sky the first thing in the morning to the way he placed his hands on his stomach when going to sleep at night. Anne and I enjoyed his ritual as play; we retain some of it in our lives, but we have not been able to accept most of it very well.

It was not possible for us to live as a couple at Sōen Rōshi's temple, so he referred us to his friend Yasutani Haku'un Rōshi, and we found ourselves once again with a teacher who did not pay much attention to ritual. He was ordained as a child but lived for the early part of his adult life as a schoolteacher and principal before he began the Zen practice that led him ultimately to become a rōshi. I can remember him saying that zazen is the important thing and that he just went along with ritual because his students wanted it.

When Yasutani Rōshi retired, Anne and I, together with the Diamond Sangha, became disciples of his successor, Yamada Kōun Rōshi. He too is not especially concerned with ritual. Generally, he leaves the dōjō procedures up to his senior disciples, who consult Sōtō priests whom they trust in order to do things in the proper way.

I recall that when I was elevated from *junshike* to *shōshike*—that is, from "associate rōshi" to "rōshi"—the ceremony was ultimately simple. I helped Yamada Rōshi bring his suitcases from our car to the guest quarters at Koko An after his arrival from the airport. He asked me to wait a moment and opened a bag, saying, "Here," as he handed me a piece of paper. I unfolded the paper, and he translated the words announcing my promotion, and then he said, "Congratulations!"

The ritual and ceremony we have in Diamond Sangha derives from our period of study with Sōen Rōshi, rather than from our other teachers, and has been quasi-monastic in its style. This was appropriate for the spirit of the early days, from 1959 to 1965. Many of our first students had been followers of occult paths, such as Theosophy and the teachings of Joel Goldsmith, and were concerned about finding a related practice.

It was also appropriate for the New Age, from 1965 to 1974, when many people joined us who wished to be full-time students of religion. The Maui Zendō flourished during these ten years in a kind of continuous training period. Since then, however, we have been in a more settled mode, and our membership has included families. We are concerned about practice for people who must necessarily spend only short periods each week at the zendō. The imperative to meet membership needs has led me to discuss lay practice and Buddhist ethics and to publish on this subject.[1] The Kahawai collective, with its focus on women's concerns, developed within our sangha. We have had communications workshops, and we are experimenting with making business decisions by consensus.

All of this is in keeping with Senzaki Sensei's conviction that American Zen must be American. However, my experiences as a teacher have prompted me to look again at ritual and ceremony as something perennial and human, rather than as an Asian import. So I began by considering the nature of makyō (uncanny realm). The Western world was introduced to the makyō phenomenon by Yasutani Rōshi in his "Introductory Lectures on Zen Practice," included in Philip Kapleau's *The Three Pillars of Zen*, first published in 1965. Yasutani Rōshi explained the term as "devil world" and quoted from the Chinese *Shūrangama Sūtra*, Keizan Jōkin's *Zazen Yojinki*, and his own experiences as a teacher to show how makyō are delusions, mainly visual, that arise during zazen and sometimes at other occasions during periods of intense practice. They can be obstructions if they are dwelt upon, but they are also signs that the student is at a crucial point in the zazen process. Yasutani Rōshi concluded his remarks on makyō by saying, "Whenever makyō appear, simply ignore them and continue sitting with all your might."[2]

When I began teaching, I soon discovered that "devil world" is justifiable as a translation etymologically, for as a term makyō relates to Māra, the incarnation of all that is evil and misleading. The experience itself, however, can be quite revealing and encouraging. I went back to the Chinese ideographs and found that "mysterious," "occult," and "uncanny" were possible translations for the ma of makyō. Thus I began using the translation "uncanny realm."

I then began looking at the content of the delusive visions that students reported and determining what they had in common. I returned to Keizan Zenji's cautions in his *Zazen Yojinki*[3] and consulted the "Fifty False States Caused by the Five Aggregates" in the Chinese *Shūrangama Sūtra*.[4] It dawned on me that while the broad classifications of delusions found in these works were useful, such psychophysiological distortions as finding the wall disappearing were listed indiscriminately with deep dreams. On listening to students, I found that some delusions were discouraging, and some were encouraging. Clearly, it was important to be more exacting about them.

My own makyō had appeared shortly before my most important realization. Seated in sesshin, working on the kōan Mu, I found myself on the stone floor of an enormous hall that had stout stone pillars extending up to an infinitely high ceiling. A line of very tall monks robed in black walked in single file around me, reciting sūtras. There was an atmosphere of remote antiquity and holiness in this dream, and I emerged from it with a strong feeling of confirmation.

Listening to students report their makyō, I recalled my own and began to understand that a lot of what Zen teachers call makyō are really inconsequential mental phenomena. I began to narrow my use of the term to the deep dream that has three characteristics: (1) a sense of the ancient, (2) a religious drama in which the dreamer is chosen or confirmed as a disciple, and (3) a sense of encouragement. The pantheon of Zen Buddhism, with its many Buddhas and Bodhisattvas, forms the storehouse for such makyō experiences. Three Diamond Sangha students—all of them women, it happens—found themselves to be Shākyamuni Buddha. One of them was covered with gold leaf; one was covered with shimmering golden light; the third was an image that was incalculably ancient. Transcending their everyday sexual identity, they found themselves at one with the deepest archetype of their religion, and all three went on to have realization at a later time.

Other students, men and women, have identified with or found themselves interacting with Bodhisattvas or old teachers, either specific figures like Kuan-yin or Chao-chou or less identifiable figures that nonetheless had Buddhist archetypal power. Still other students experienced more unspecific dramas set in ancient times but always with themselves as central actors in a deeply moving ritual.

The profound sense that "something important is happening to me," which accompanies makyō, and the fact that realization seems to be made possible by the experience combine to convince me that it is important for us as Zen students to cultivate the dimension of makyō through ritual and ceremony. Or, to take this notion a step further, to identify ritual and ceremony as enacted makyō. An even further step would be to experience our rituals and ceremonies as those of Shākyamuni and his disciples and those of all his succeeding masters and their disciples.

Thus our ceremonies can establish our zendō as the Bodhi-manda, the sacred dōjō of the Buddha, in which Maudgalyāyana, Mahākāshyapa, Shāriputra, Ānanda, and the other ancient worthies dance in attendance. Make no mistake—they dance and sit and walk with us here. Just as the experience of makyō seems to be a precursor of realization, so the ancient scene of sūtras and bows enacted in our zendō can set the scene for the Buddha's own experience as our own.

Notice that Wu-men describes how you walk hand in hand with all the ancestral teachers in our lineage, "the hair of your eyebrows entangled with theirs, seeing with the same eyes, hearing with the same ears."[5] This is makyō, it seems to me, and these words precede his description of realization.

There are makyō in our kōan study as well. Have a look at the story of Yang-shan visiting the hall of Maitreya in Case Twenty-five of *The Gateless Barrier* or Wu-cho conversing with Mañjushrī in

Case Thirty-five of *The Blue Cliff Record*. These are real, historical people interacting in an uncanny realm with mythological archetypes. The story of Yang-shan at Maitreya's hall begins, "Yang-shan dreamed he went to Maitreya's place." Yüan-wu, in his comment on Wu-cho's extended discourse with Mañjushrī and later with his boy attendant, describes the scene as illusory.[6] We take up the makyō of Yang-shan and Wu-cho as kōans.

I believe that while the old teachers are right, we should not dwell on makyō or any other delusions. Still, we can use them, even encourage them as upāya (skillful means). In the Buddha's own dōjō we offer flowers, water, candles, incense. This is our temple of the Buddha, this building, this sangha, this body. The Buddha can perceive the Morning Star from here.

1985

Kōans and Their Study

Kōans are tiny doors that open to great vistas, inviting us to wander through endless gardens. They are the folk stories of Zen. Like folk stories, their expression is presentational, rather than discursive, to use Susanne Langer's terminology.[1] They are poetical — and sometimes, though not usually, they can be nonverbal. In any case, they are not explanatory.

R. H. Blyth said many times that Zen is poetry.[2] Poetry too — and I distinguish Shakespeare from Joyce Kilmer here[3] — has endless scope. Like Chü-ti's finger, upraised in response to every question, it can never be used up.[4] In the realm of haiku, seventeen syllables have profound implications. Bashō wrote:

> The little horse ambles clop-clop
> across the summer moor —
> I find myself in a picture.[5]

Bashō's disciple Sampū painted a picture of Bashō nodding along on his little horse, completely absorbed — subjective and objective fallen away, the inside world enlarged to fill the summer moor; the summer moor filling the inside world.[6] We fall away with Sampū falling away with Bashō, who fell away with the rhythmic clop-clop and the warm summer vista. Everything vanishes in a single, unified, subtle experience of many dimensions.

Like haiku, the brief kōan opens to broad prospects. Moreover, both haiku and the kōan evoke the perennial, for in many cultures and times we find the same fascination with the tiny sound, scent, taste, sensation, thought, or thing that is potent with expansive possibilities. Robert Pinsky, for example, comments quite topically, "Successful computer entertainments in language have tended to be about the way something quite small and unitary—the chip, the microelectronic impulse, the bytes, the little gray box on one's desk—opens up into something very large and elaborate. This opening up, the discovery of much in little, seems to be a fundamental resonance of human intelligence."[7]

Everywhere this resonance appears. As I step outdoors, suddenly a thrush bursts into song in the little milo tree by our front porch. House, garden, memories, and plans completely disappear, and in the incredible silence, she evokes from my heart her lovely voice again and then again. This is the "entry into the inconceivable" of song within song, life within life. It is the experience described metaphysically in the *Hua-yen Sūtra*, where the little door to Maitreya's tower opens for Sudhāna to a vast panorama of other towers that in turn hold inner panoramas of innumerably more towers.[8]

Each kōan is a single experience like Sudhāna's, a door that opens to the myriad things in their dynamic interaction—to the myriad things as all things—to the myriad things as nothing at all—but from just a single vantage. Yün-men said, "This staff has become a dragon. It has swallowed up the whole universe."[9] This passage is but a part of one case in *The Blue Cliff Record*, a book of one hundred kōans, giving one hundred vantages. When they are enriched with insightful comments and poems, then you have ten thousand vantages.

Though it may seem complete, the experience of a single kōan is simply a new beginning.[10] When Yamada Kōun Rōshi experienced the mouth of the universe opening to laugh with his own great laugh, he realized the mind is truly "the mountains, the rivers, and the great Earth, the sun, the moon, and the stars."[11] But then he took up checking questions in Yasutani Rōshi's room, the introductory kōans, and the ensuing hundreds of kōans of formal study in the Sanbō Kyōdan. After that he deepened his insight, on and on, as we found, sometimes to our surprise, in his kōan review seminars.

Yüan-wu likened Yün-men, holding forth his staff before his assembly, to the Buddha earlier twirling a flower before *his* disciples.[12]

With Yün-men's staff and the Buddha's flower, the tao of the kōan is revealed, among other things. Mahākāshyapa's smile in response to the twirling flower initiated a transmission that has not yet come to an end. Thus the process of enrichment goes on and on.

"Do you understand this specially transmitted mind?" Yüan-wu asks. A tiny particular encloses everything in its own way. Unless you enter such a particular, unless a particular enters you, you are lost in sights, sounds, tastes, scents, and sensations, with the end of their sequence looming like darkness that gradually subdues the daylight.

Everybody knows about this darkness. Everybody knows you can't take sights and sounds and sensations with you when it descends. With this understanding comes anguish, which, as the Buddha said, is everywhere. Under the Bodhi tree the Buddha found his compassionate imperative confirmed, and he was ready to show the way out of anguish.

All our great ancestors were motivated in this way. From the beginning their vow was to enable beings to cross to the other shore. Their path is only clear in the vast silence of no-space and no-time that opens with a glimpse of the Morning Star, with the master holding up a staff, or with the song of the thrush under the overcast sky.

The *Heart Sūtra* says that Avalokiteshvara, practicing the deepest spiritual wisdom, clearly saw that all perceptions and things perceived are essentially empty and that this kind of realization transforms anguish and distress.[13] The emptiness of the *Heart Sūtra* is the universal—the absolute, if you will—full of possibilities but void of any characteristic.

Once you uncover this void, where feelings and thoughts and their objects are without substance, you experience enormous relief. "It" doesn't matter anymore. One such realization is the beginning of kōan study. It is one fragment of the holographic plate. Many fragments must be put together in a lifetime of practice.

The step-by-step process through one's own ox-herding pictures begins with the star or the staff or the birdsong in emptiness, as emptiness itself—the fundamental complementarity of our universe.[14] This complementarity is confirmed at every step of the way.

There are many kinds of complementarity facing the human being: male and female, for example, or giving and taking. However, the most intractably separate double stars of the mind are the particular and the universal, form and emptiness. This is the realm of the kōan. We sense the dynamic unity of the distinctly different elements, but with our inherent, either-or patterns of thinking we find no connection.

Something has to give. Either kōan study must go, or the path of reason. Thus, early on, the student who elects to pursue the path of Zen Buddhism gives up history and philosophy as basic tools and takes up the way of poetry. The way of poetry is the way of staring at the word or words with only the question "What is it?" occupying the mind. The point either emerges or it doesn't. If it doesn't, one's only recourse is to go on staring. A sound practice of zazen, of Zen meditation, is essential. It is essential also to have a good teacher to light the path.

The temptation to philosophize is very strong, however. Some old teachers gathered kōans into types: the Dharmakāya type, the Nantō type, and so on, which you can study in *Zen Dust*.[15] I don't find these categories especially useful, except as hints of the variety that can be found in the practice. The teacher who dwells upon these categories is like the one who dwells upon samādhi power. At two ends of the same spectrum that ranges from the brain to the large intestine, these teachers of reason or physical coordination do not dream of the Buddha's world or of Yün-men's world. They cannot bring our great ancestors to life. They expose their limitations clearly, particularly in their *mondō*, their formal public encounters with their students.

The mondō tells all. One gets the impression on reading some studies of Zen that the mondō is a kind of give-and-take, a way of striking sparks. I myself have suggested as much in the past. Not incorrect, but now I find the mondō better described as a drama or a dance in which the players or dancers bring forth the dynamic being of the Dharma. "Give-and-take" is a static kind of description for the joyous movement and the unexpected sally that the enlightened mondō can provide. See, for example in *The Blue Cliff Record*, Wuchiu dancing with a monk, swapping blows of the staff and the staff itself, all the while bantering back and forth about giving and receiving.

A monk came to Wu-chiu from the assembly of Ting-chou. Wu-chiu asked, "How does the Dharma of Ting-chou differ from ours?"

The monk said, "It does not differ."

Wu-chiu said, "If it does not differ, then go back there," and gave him a blow with his stick.

The monk said, "Your stick has eyes. You shouldn't hit a fellow wantonly."

Wu-chiu said, "Today I've hit one," and gave him three more blows. The monk went out.

Wu-chiu called after him, saying, "All along, there was someone receiving it."

The monk turned and said, "What can I do? Your Reverence is holding the stick."

Wu-chiu said, "If you like, I'll hand it over."

The monk came nearer and snatched the stick from Wu-chiu's hands and gave him three blows. Wu-chiu said, "Blind stick! Blind stick!"

The monk said, "There's someone receiving it."

Wu-chiu, "It's a pity to beat a fellow wantonly."

The monk bowed. Wu-chiu said, "Yet you act this way."

The monk laughed loudly and went out. Wu-chiu said, "That's all it comes to; that's all it comes to."[16]

Only the well-trained can dance; only the inspired can inspire others; only the realized ones can see through each other.

Hinduism provides us with the image of Shiva dancing. His one foot is raised—when he lowers it, the world ends. This is the dance of life, which includes death. We dance creation and annihilation too, but our dance also includes the not-born and not-destroyed with the life that continually passes away.

Mahayana Buddhism provides the model of the Net of Indra, where each point perfectly reflects and contains every other point.[17] While the Net of Indra might seem a rather static model, our Shiva gracefully dances and chants reflection and containment with each "Good morning, how are you?"

Recently I have been going through Case Forty-eight of *The Gateless Barrier* with a student. It goes like this:

monk Wai-tao, whom he supported financially for many years, and Dr. Suzuki. Much of the time he spent in Asia during the later years was devoted to consulting with his resource people about their translations.

19. Kerouac, *The Dharma Bums*, 202.
20. "Inside the FZI, 3," *Zen Notes* 28, no. 4 (April 1981): 2.

THE BRAHMA VIHĀRAS

1. Walpola Rahula, *What the Buddha Taught* (New York: Grove Press, 1959), 75.
2. Robert Aitken, *The Practice of Perfection: The Pāramitās from a Zen Buddhist Perspective* (New York and San Francisco: Pantheon Books, 1994), 31.
3. Cf. Thomas Cleary, trans., *Book of Serenity* (Hudson, N.Y.: Lindisfarne Press, 1990), 17.
4. *The Gateless Barrier: The Wu-men kuan (Mumonkan)*, trans. with commentaries by Robert Aitken (San Francisco: North Point Press, 1990), 7.
5. Aitken, *The Gateless Barrier*, 9.
6. Gary Snyder, *Earth House Hold* (New York: New Directions, 1969), 90–93.

EMMEI JIKKU KANNON GYŌ

1. In this essay I use the familiar Sino-Japanese with reference to the sūtra and the Chinese with reference to Chinese names.
2. Japanese scholars have attempted to reconstruct the originals from their transliterated forms. See, for example, D. T. Suzuki, *Manual of Zen Buddhism* (New York: Grove Press, 1960), 21–23.
3. For a discussion of the origin of the sūtra, see Philip B. Yampolsky, trans., *The Zen Master Hakuin: Selected Writings* (New York: Columbia University Press, 1971), 185.
4. Yampolsky, *The Zen Master Hakuin*, 18–24, 185–86. See also Barbara E. Reed, "The Gender Symbolism of Kuan-yin Bodhisattva," in *Buddhism, Sexuality, and Gender*, ed. by José Ignacio Cabezón (Albany: State University of New York Press, 1992), 159–80.
5. Haku'un Yasutani, "Theory and Practice of Zazen," *The Three Pillars of Zen*, ed. by Philip Kapleau (Boston: Beacon Press, 1965), 27–28; also, Isshū Miura and Ruth Fuller Sasaki, *Zen Dust: The History of the Koan and Koan Study in Rinzai (Lin-chi) Zen* (New York: Harcourt Brace & World, 1966), 292.

A monk asked the priest Kan-feng, " 'Bhagavats [Buddhas] in the ten directions; one straight road to nirvana.' I wonder where that road is."

Kan-feng lifted up his staff, drew a line in the air, and said, "Here it is."

Later the monk asked Yün-men about this. Yün-men held up his fan and said, "This fan jumps up to the Heaven of the Thirty-three and strikes the nose of the deity Shakradevendra. Give a carp of the Eastern Sea one blow, and the rain comes down in torrents."[18]

There are three dancers here: the monk, Kan-feng, and Yün-men. Yün-men has two parts: the first his response to the monk and the second his final comment. Kan-feng and Yün-men were good friends whose interactions appear elsewhere in our study.[19] Wu-men adds choreography of his own:

One goes deep, deep to the bottom of the sea, and winnows the mud and pumps up the sand. The other goes high, high to the top of the mountain, and raises foaming waves that spread over the entire sky. Maintaining, releasing, each using but one hand, they safeguard the vehicle of the Tao. They are like two children, running from different directions, who collide with each other.[20]

Notice that Kan-feng presents the road to nirvana; he does not merely point to it. Yün-men presents the road too, in a delightfully childlike way. How do you dance these presentations?

More than thirty years ago, when I was a leader at the Koko An Zendō but not yet started on my kōan study, we had annual sesshins with Yasutani Haku'un Rōshi. With his inspiration, several people began to move in their practice. I was the monitor of the interview line, among other things, and during the interviews I sat in the alcove of the Zen hall with the front door open, keeping track of the coming and going of students. The rōshi met with them in the cottage in the front garden, and I could see through the screens. Sometimes, there before the rōshi, students would stand up and seem to walk around. "What's going on?" I wondered. Well, they were dancing, winnowing the mud, raising foaming waves.

Yün-men remarks finally, "Give the carp of the Eastern Sea one blow, and the rain comes down in torrents." This is how the dance works in the natural world. How would you show Yün-men's intention here?

In response to this question, you are not necessarily called on to present an elegant metaphor; in fact, elegance can be a pitfall. In the Sung period, Ta-hui was so alarmed by the tendency to dwell on the beautiful phrases of *The Blue Cliff Record* that he burned its printing blocks, almost destroying the book forever.[21] I think his message is that the experience of Mu itself, for example, is not affective or cognitive or aesthetic. However, if it is true and clear, *then* it can be danced. It can be felt and expressed.

Another pitfall is to use kōans superficially to skip through the dark side of human nature. If you are truly engaged in the study, you will find that the way of Zen is not merely a metaphysical exercise and not merely sweetness and light. The analogy of kōans and folk stories is once again instructive. Folk stories, as Heinrich Zimmer tells us, can open their tiny doors to evil and the arduous path to conquer it.[22] You will find classical Zen Buddhist cases that take you along on this same difficult but rewarding path. See, for example, Yün-men's words, "Every day is a good day."[23] Yün-men leads us through self-betrayal, through the malice of others, through the karma created by awful events that are tied intimately with our past, and finally through a Pollyanna kind of denial—until at last we find the original dwelling place of our wise yet playful master, right here in our polluted world.

I think of a friend, a school counselor who has never practiced zazen. He jives with his students and they love him for it, but his jiving is different from theirs. It is always elevated just a bit. They come up to his level, and then he elevates just a bit more. This is the perennial way of teaching, the counselor among public school children, Chaochou among all of us.

I think of another Bodhisattva friend, a copy editor of a metropolitan newspaper. Casting about for ways to apply his Zen Buddhist practice in his daily life, he decided, "I'm going to bring harmony to my workplace." This was no casual undertaking. The usual newspaper office is loaded with jousting egos. Yet with perseverance, he could acknowledge in his modest way, after a year or so, that he had been effective. The office was much more harmonious, simply because he practiced keeping cool in crises, silent in the midst of gossip, agreeable in the heat of conflict. By his simple presence, and perhaps with just a word here and there, he elevated the jiving. By his simple presence, and with just a single world, Chao-chou elevated generations of gossipy monks and laypeople.

Recall the people who influenced your life in a positive way. Probably they were ones who could say, "Yes, I see your point," even when they occupied a very different ground. With this you felt included, and there was space for discussion and reconciliation. In Japan, you hear the word "yes" a hundred times more often than you will hear the word "no." You will hear "Is that so!" again and again when it clearly isn't.

Thus Japan is a nation of mediation and consensus. Sometimes this mediation becomes self-centered manipulation—that's the dark side. From time to time in the sangha too, we can see how a sham kind of inclusion can be exploitative. Interest in others is confined to scandal and manipulation. The teachings of Chao-chou and his kinfolk are forgotten. Common decency is forgotten.

Those old teachings and ideals of common decency can devolve into rules and regulations, but shining through commandments and precepts is the conscious or unconscious knowledge of inclusion. The family is the model. The sangha is the model. We are clearly responsible for taking care of each other.

In the dōjō of the individual self are many students on one path, within one Dharma. When this becomes clear, then by tone and manner, or perhaps in a personal story, the next stepping-stone is revealed.

I think of Patrick Hawk, my Dharma heir in Amarillo, as one who teaches by manner. I once found myself in his quarters, awaiting his return from an errand. He had just two rooms—an outer reception room and an inner bedroom. He kept his reception room like a little temple. What was his bedroom like? I peeked, and there was another little temple. I had to look closely to see the bed, folded up neatly and unobtrusively in a corner.

Yet I don't think I have ever heard Pat Hawk cite the Japanese proverb "Clean as a Zen temple." He doesn't need to. Living with his students, his teaching is his person. I sometimes cite the proverb, but I'm afraid that my words are hollow. I am saying, in effect, "Do as I say, not as I do," for my own quarters tend to be pretty messy. "Do as I say" doesn't include anybody. Neither does the response "Practice what you preach."

In a recent conversation, Maha Goshananda remarked to me, "We are forest monks, but there are no more trees in Cambodia." He was saying this as the supreme patriarch of Cambodia, so his words were not only potent with pathos but also with determination to find the way of monks bereft of their traditional habitat. So he leads peace walks along roads littered with land mines, through villages and fields that are daily torn by civil war, calmly and serenely serving as the model and archetype of the Buddha's way of infinite compassion.

Maha Goshananda casts light on the way of inclusion and compassion, Patrick Hawk casts light, my friends the counselor and the editor cast light. Inclusion is compassion, suffering with others, and with compassion teaching and guidance arise. Kōan study can be a door to this kind of work. A single kōan illumines many kinds of inclusion. The single word of Chao-chou shines forth most instructively. When we make that light our own, we ourselves shine forth.

My first teishō (Dharma talk) in each sesshin is devoted to the kōan Mu.

> A monk asked Chao-chou, "Has the dog Buddha Nature, or not?"
> Chao-chou said, "*Mu.*"[24]

Chao-chou, in his venerable age and profound wisdom, comes forth as the Tathāgata, as Buddha Nature itself, all of a piece with his Mu. He shows the very point of the question. He is the point and shows it as himself. He also includes the monk in his presentation, confirming the doubts the monk had implied about Buddha Nature. By responding "Mu—no, does not have," he is saying, "You are right. For sure, the dog doesn't have Buddha Nature." You and I are on the same ferryboat here. We are moving along in the dynamic space of sangha, an organic being in which each of us is a body part. I confirm your words as my own.

Nonetheless, this agreement points to another step. "By the way," he is implying, "what is this 'does not have'? What is Mu?"

Chao-chou's Mu is, of course, not discursive at all, but it is open to discursive explication: Buddha Nature is essential nature, sometimes rendered shūnyatā, or the void. Yamada Kōun Rōshi used the model of a fraction to symbolize reality, with the Greek letter *alpha* (α), representing all phenomena, as the numerator and the mathematical symbol for infinity (∞), surrounded by a circle, representing empty infinity, as the denominator. Phenomena are denominated by empty infinity—infinite emptiness.

Any fraction equals a number, so what is on the other side of the "equals" mark? The fraction of phenomena over emptiness is equal to our reality, of course—people, animals, trees, towers, streets—things as they are. The rōshi's fraction is useful in clarifying the infinite emptiness of the plenum and of each of its elements. The *Heart Sūtra*, too, clarifies the identity of emptiness and the world of form.[25]

Old grandmothers like Yamada Rōshi and tattered texts like the *Heart Sūtra* can set the tone and weave the backdrop for students in the lecture hall, but in the dokusan room you meet the same master squatting like a dragon in his lair, the texts long since ripped to smithereens. This is the cave where reason is suicide, and the baker simply announces the price of muffins, with no exegesis.

Here is the illumination of Hsiang-lin:

A monk asked Hsiang-lin, "What is the meaning of Bodhidharma's coming from the West?"
Hsiang-lin said, "I'm stiff from sitting so long."[26]

Students of Zen Buddhism can relate to Hsiang-lin's stiffness. After long days of zazen in sesshin, people rise slowly, even painfully, for kinhin, the formal walk between periods of sitting. But how many can make any connection between their stiffness and pain and the essential teachings of Bodhidharma? That is the edge of Hsiang-lin's response—the elevation. After all, Bodhidharma is to blame for this "jungle of monks at sixes and sevens,"[27] this confusion of bowing and squatting and reciting old sūtras that we call a training period.

When people come to me with a response to their kōan—for example, Hsiang-lin's answer to a monk's question about Bodhidharma's imperative—commonly they begin by saying, "Well, I think he . . . " and that's where I intervene. "Well, I think he . . . " puts Hsiang-lin or whomever halfway across the world and back in time some thousand years. Not very intimate. Not very inclusive.

History and geography have their uses, sometimes even in the dokusan room, when the teacher might comment on a cultural aspect of a case once the case itself is resolved. Such cultural overtones are, however, relatively uncommon. Most cases deal with basic human matters that are pretty much the same for old Chinese and modern North Americans, Australasians, and Europeans. Everybody can relate to Hsiang-lin, grimacing as he lifts his feet from his lap.

Over and over we experience such intimacy in our daily lives. Around the water fountain we hear about somebody's mother who has Alzheimer's disease. We feel an upwelling of empathy. Though it is the kind of encounter that happens frequently, it can each time be an enlarging experience. It can be transformational to the degree that we are open to experience it and willing to be guided by it.

Taken by themselves, Hsiang-lin's words are far more casual than agonized sharing around the water fountain. In context, however, like Chao-chou's Mu, they bring together in dynamic unity the past, present, and future, near and far, seen and unseen, the void itself with the riches of the world. We add our own notes to this resonance in the profound intimacy of even longer ago and farther away and, indeed, of no-time and no-space. It is the chance for truly profound transformation, again depending on how willing we are to be guided by such experiences.

In our daily lives, the family, the sangha, the workplace are the laboratories of intimacy. In our dōjō even etymology can be one of our teachers. It is interesting to note that the suffix meaning "monk" in Japanese is *sō*, which means sangha. It is used in the compound *Zensō*, "Zen monk." The individual monk is the sangha, and the same is true for each of us. The individual student is the Sangha Treasure, and though as laywomen and laymen we have other identities, as students of Zen Buddhism it becomes clear that we are the constituents of one body that is beyond any sectarian designation.

It is easy to forget the penetrating truths of Chao-chou and Hsiang-lin. It is easy to forget how they include us in their wisdom, how we include them and indeed all beings. We fall into gossip. We triangulate, divide off into exclusive subgroups, or just forget to say hello. Sometimes severe misunderstandings burst forth and people pick up and leave in anger. If we can't get along in our little practice center, how can we expect global society to make it?

We are in it together, this ferry to the other shore. If I take my role seriously as an oarsman—if I am steady and don't move around a lot—then others will surely pull their oars and be steady as well, and we'll get there as an organic community.

1995

Marriage as Sangha

A Talk

this little bride & groom are
standing)in a kind
of crown he dressed
in black candy she

veiled with candy white
carrying a bouquet of
pretend flowers this
candy crown with this candy

little bride & little
groom in it kind of stands on
a thin ring which stands on a much
less thin very much more

big & kinder of ring & which
kinder of stands on a
much more than very much
biggest & thickest & kindest

of ring & all one two three rings
are cake & everything is protected by
cellophane against anything(because
nothing really exists[1]

"Nothing really exists." Here the poet, e. e. cummings, presents the fundamental, perennial human problem, which the Buddha defined as duhkha, the anguish we experience when we glimpse our nonbeing. Duhkha is also the way we disguise our anguish and protect our disguise—in cummings's metaphor, with ideals made of candy and cake, wrapped in airtight cellophane.

The "nothing" that cummings faced with such good humor is the emptiness set forth in Buddhism—the vacancy that is your nature and mine and the nature of the universe. It is not vacuum, for it contains all things, as the sky contains the stars. It is not chaos, and yet it is shot through with unreliable, dangerous asteroids. It is harmony—not merely the harmony of candy with cake but of death with life, of the unknown with the known. And the dynamics of harmony involve the particularity of each being and the infinite variety of all beings.

The Buddha and his successors saw clearly into the interaction of the many beings as they made their way through the unknown. To fulfill our potential as human beings, he said in effect, we strive for nobility, and further, we establish our path of nobility with our vows. We give our word that we will be decent to one another and give each other support. In this way we turn the wheel of the Dharma for harmony and enlightenment in the world.

We give our word in the wedding ceremony, but we Buddhists in the West who inherit the Dharma from Japan find that such a ceremony is not a central part of our tradition. Most of our Japanese Buddhist ancestors were married in Shintō services, and what Buddhist wedding ceremonies there are tend to be buried in archives somewhere. Western Buddhist leaders have consulted those old transcripts, but since they were so seldom used, they haven't felt bound by them. Thus, Western Buddhist wedding services are a relatively new tradition, and there are marked variations from sect to sect. However, all of them—without exception, I am sure—use ancient elements of other Buddhist rituals in their forms.

The wedding ceremony we use in the Diamond Sangha is based on a manuscript I developed with the Venerable Eijō Ikenaga, priest of the Honolulu Myōhōji Mission. Early in his ministry, he was called upon to do a ceremony in English, so he wrote one out in Japanese, and the two of us translated it. Working with many couples since then, I have revised this manuscript dozens of times, but the basic form is the same: vows by the bride and groom to follow the Five Precepts of the lay Buddhist.

This set of precepts, called the Pañcha Shīla, is surely among the most ancient of vows; they are based on the pre-Buddhist ideal of ahimsā, or nonharming. In the Mahayana context, the Pañcha Shīla is not merely a set of five promises to avoid negative conduct but it is also a positive affirmation of the path of nobility. So in the Diamond Sangha wedding ceremony, the vow not to kill becomes your promise to help and encourage your spouse in a most generous way; the vow not to steal becomes your promise to respect the unique thoughts of your spouse as expressions of someone on the path of maturity; the vow not to misuse sex becomes your promise to give of yourself fully in your love; the vow not to lie becomes your promise to be faithful and true; and the vow not to indulge in intoxicants becomes your promise to keep your mind and all the circumstances of your marriage clear.

Some couples also want to include promises from the *Book of Common Prayer* to take each other through joys and sorrows, sickness and health, and so on, until they are parted by death. This is fine with me. Give your word that you will encourage your spouse and be faithful, and your wedding ceremony will be a beacon for your pilgrimage together, and your pilgrimage in turn will be a beacon for the paths of others.

Wendell Berry says in his essay "Poetry and Marriage":

> The meaning of marriage begins in the giving of words. We cannot join ourselves to one another without giving our word. And this must be an unconditional giving, for in joining ourselves to one another, we join ourselves to the unknown. We can join one another *only* by joining the unknown. We must not be misled by the procedures of experimental thought: in life, in the world, we are never given two known results to choose between, but only *one* result that we choose without knowing what it is.[2]

This is the Way of Marriage through the shadows of doubt. We are not given two known results to choose between; for example, we are not given divorce and marriage as two options, but just the one result that is constantly unfolding. If this unfolding leads irrevocably to divorce, then that is the result that could not have been foreseen. Our commitment is to making our way through the unknown together, whatever happens.

The candy bride and groom wrapped in cellophane disguise the noble path of making our way through the unknown. We are not only creating a family and making it possible to have children and to bring them up safely, we are walking the path of personal growth together.[3] Facing the unknown, treading the unknown, we encourage each other through the dark night that inevitably sets in. If we try to dismiss this dark night as a "midlife crisis," we find it cannot be dismissed and must be lived through. Perhaps our path will be Buddhist, perhaps Christian, perhaps Judaic, perhaps humanistic with no religious name. In any case, in the perennial sense, it is a religious path. It is the way of wholeness and wellness we choose to take together.

There is another point. In creating a marriage, the bride and groom create a new being. As wife and husband, and then possibly with children, the family is a sangha with its own needs, its own delights and suffering. It is a tangible entity, separate and yet inclusive, in which the family members take part and within which they fulfill their individual lives.

Thus our wedding vows, the words we stand by in our marriage, are not only promises to one another but to our marriage itself, the tender being we bring into existence and which we nurture with our honorable conduct and love. In our promises to be true to each other, the words we use are the ancestral jewels of our parents, our grandparents, our great-grandparents, and so on back into the misty past. Keeping those jewels faithfully, we nurture the being of marriage through its own natural life.

When it is done, and the husband or wife has passed on, the other will linger for a while, savoring a marriage that still continues, the life that turned the cosmic Dharma wheel a little and brought the innate harmony of all beings a little closer to fulfillment.

1986

Death: A Zen Buddhist Perspective

Death is treated as a teaching in Zen Buddhism. It reveals and enriches the truths of impermanence, compassion, and interdependence. As a metaphor it reveals the nature of step-by-step practice and of realization.

Zen teachers of the past were commonly able to foresee their own deaths, to prepare for them, and to find a dignified and appropriate way of dying. Daiō Kokushi, the de facto founder of Rinzai Zen in Japan, announced the date of his death a year in advance to the day.[1] His grandson in the Dharma, Kanzan Kokushi, made his own dying a ritual:

> On the day of his death Kanzan entrusted his affairs to his sole Dharma heir and dressed himself in his traveling clothes. Then he went out from the abbot's quarters and, standing alone beside the "Wind and Water Pond" at the front gate of the temple, he passed away.[2]

Hung-chih, editor of the *Book of Serenity*, traveled around saying good-bye:

> One day in the autumn of 1157, when he was in his sixty-seventh year, Hung-chih put on his traveling garb and jour-neyed down the mountain for the first time in nearly thirty years. He visited the commander of the army, the government officials in the district, and the patrons of the temple, thank-

ing them all for their kindness during the years and saying good-bye. On the tenth of November, the master returned to the temple. The following morning, after bathing and changing his robes, he sat down in the formal position and gave a farewell talk to his assembled disciples.

Then he asked his attendant to bring him writing materials. He composed his death poem and passed away with his brush in his hand.[3]

There is implicit teaching in all of this, but the old teachers also used death explicitly to guide their disciples and the rest of us. Yüehshan called out in a loud voice one day, "The Dharma hall is falling down! The Dharma hall is falling down!" The monks rushed to hold up the pillars. He clapped his hands and laughed loudly, saying, "You don't understand." He then passed away.[4]

There are many such stories of death used as an upāya, a skillful means of turning the Dharma wheel. Death poems were upāya. Here is Hung-chih's composition:

> Illusory dreams, phantom flowers—
> sixty-seven years.
> A white bird vanishes in the mist,
> autumn waters merge with the sky.[5]

Bassui advises a dying man:

> If you think of nothing, wish for nothing, want to understand nothing, cling to nothing, and only ask yourself, "What is the true substance of the Mind of this one who is now suffering?" Ending your days like clouds fading in the sky, you will eventually be freed from your painful bondage to endless change.[6]

Bassui is advising his student to persist quietly and calmly with his kōan to the end, promising him liberation. But what happens when the white bird vanishes and the clouds fade? We turn to etymology, that wise educator, and find that the word *senge* in Japanese, the term used for a death of a Buddhist master, means "to pass into transformation." While Bassui and Hung-chih could speak of vanishing or fading, death can also be considered a becoming.

In *The Gateless Barrier*, I quote Yamada Kōun Rōshi asking a student, "What do you think of death?" The student replied, "Why, it's like when a bus stops before you—you get on and go."[7] This student was my wife Anne, who took her name out of the story when she helped to edit the book. Her death many years after that dialogue made clear how deeply committed she was to the truth of her words.

Anne had suffered a massive heart attack and was breathing heavily. The doctor had come into her hospital room and was questioning me about her living will. Did I agree that there should be no intervention to prolong her life artificially? I agreed that I wanted the process to be natural. I was watching her as I spoke. Her breath became quieter and a look of the utmost determination came over her face. She pressed her lips together, her pulse subsided, and she passed away.

It seems that she could hear me speak and that she took my words supportively. The bus had come, so she took the appropriate action. She stepped aboard and sat down with Kanzan, Hung-chih, and so many other ancestors who died showing the rest of us the Way.

I am resolved to learn from Anne's readiness, which is *all*, as the prince says in *Hamlet*. This is a matter of preparing oneself. Toushuai said, "When you are freed from birth and death, you will know where to go. When your elements scatter, where do you go?"[8] This is an ultimate kind of kōan. Understanding it involves cutting your bondage to the endless fluctuation—cutting your attachment to the sequence of your movie and finding your home in its particular frames.

In each frame, the metaphor of death offers a handle to the practice. Dōgen Kigen Zenji places death among the countless acts of dāna (giving) that make up our daily work:

> When one learns well, being born and dying are both giving. All productive labor is fundamentally giving. Entrusting flowers to the wind, birds to the season, also must be meritorious acts of giving.[9]

In this passage, Dōgen Zenji implies that death (and birth too, in a different way) is more than relinquishment, even more than giving. It is entrusting. When we are practicing—when we are turning the Dharma wheel—all our acts are those of entrusting. At the Pālolo

Zen Center we entrust our cat to keep rats from the premises, but we also entrust him to keep birds away. Entrusting makes the world go around. Kanzan entrusted his work to his sole Dharma heir. On his deathbed, Lin-chi entrusted his Dharma to San-sheng.[10] Along with the mystery of death comes entrusting one's work to the world.

This means entrusting to the future, of course. I once asked Yamada Rōshi about life after death, and he replied, "Well, there is always the karmic side." Indeed. Karma is action. There is individual karma, social karma, world karma, universal karma. The specifics of these karmas go on and on, impelled from the past, absorbing influences from each other, unfolding into the future. What would be my specific and what would be yours, going on and on, ever changing?

I don't know, and I take my cue from the Buddha's unwillingness to conjecture about such things. I don't even know about my specific in this life very well. Coming to terms with it is my lifetime task and involves facing perennial questions. "Who am I?" This is the basic query, and while in so many words it might have worked as a kōan for Ramana Maharshi, it tends to take the rest of us around and around in our cortex. "Who is hearing that sound?" asked Bassui Tokushō Zenji.[11] That works better for most people, for the self dies in the process. "What is my task?" is a useful prompt that helps to clarify dying as daily practice. You forget everything as you greet friends or water the plants. In the nonce of dying, you forget everything, bequeathing your concerns to family, friends, and all beings.

If you can't visualize this kind of release, then you have two options: (1) a religion that promises eternal life or (2) no religion, which can be the condition of despair. Here is Philip Larkin's despair in his poem "Aubade":

> I work all day, and get half-drunk at night.
> Waking at four to soundless dark, I stare.
> In time the curtain-edges will grow light.
> Till then I see what's really always there:
> Unresting death, a whole day nearer now,
> Making all thought impossible but how

And where and when I shall myself die.
Arid interrogation: yet the dread
Of dying, and being dead,
Flashes afresh to hold and horrify.[12]

In our heart of hearts, we all know about this horror. Katsuki Sekida, our first resident teacher at the Diamond Sangha, used to tell us that when he was a child, falling asleep in the comfort of his bed in his happy home, he would suddenly hear his own voice, crying out in terrible tones, "You must die!" He would call to his parents and sob in their arms, unable to explain his anguish, his duhkha. As a child he was not asking for help to carry him through the night but for assurance that the night was not there. As an adult, however, he and all other worthy students of religion have sought their way right through their terror, not around it. They shun the teachers who devise ways to avoid terror. As Simone Weil warned, "Religion, insofar as it is consolation, is a hindrance to true faith."[13]

The *Heart Sūtra* assures us, "Bodhisattvas live by Prajñā Pāramitā with no hindrance in the mind; no hindrance, therefore no fear."[14] The word we translate as "fear" is really "terror"—Larkin's terror, little Katsuki's terror, human terror. The English word "fear" is easier to chant in that context, but we mustn't neglect the true meaning. Hakuin Ekaku Zenji asks, "From dark path to dark path we wander; when shall we be freed from birth and death?"[15] When shall we be freed from our terror?

Dōgen Zenji's father died when he was two years old, and when he was seven his mother died. He recalled how watching the smoke from the funeral pyre at his mother's funeral impressed him deeply and sorrowfully with the transience of life.[16] Throughout his career of teaching he linked this awareness of transiency with bodhichitta—the desire for realization, the desire for enlightenment, the imperative for realizing the Buddha.

Bodhichitta is what distinguishes Buddhism and Zen from world religions generally. The pilgrim looks directly into the fact of death, into the fact of impermanence, and finds there the solace that others find in the notion of heaven and eternal life. What is that solace? Haha! How truly beautiful everything is!

We find maturity on this path in the death poems of Buddhist teachers and haiku poets, collected in a recent anthology. Here is one of them, by the nineteenth-century poet Bokkei:

> Oh, cuckoo,
> I too spit blood—
> my thoughts.[17]

The cuckoo shows its red mouth when it sings. My thoughts are red like blood, Bokkei is saying. My dying is like the welling up of thoughts, the song of birds—an extraordinary expression of inter-being.

Compare Bokkei's presentation with the famous haiku by Issa, on the death of his baby daughter:

> The dewdrop world
> is the dewdrop world,
> and yet—and yet.[18]

"It is true that this world is transitory," Issa is saying. "All beings are ephemeral. I know this, but when I am faced with the death of my baby girl, I look desperately for something to give me hope and comfort." This is the natural, human way of dealing with anguish, to treat it as an event that was brought forth by implacable exterior circumstances.

Natural—but, Issa, you are not addressing death squarely. There in your grief itself is your emancipation. Your tears are the blood of the universe, coming forth elsewhere in the song of the cuckoo and the darting of geckos. Each breath is truly inspiration and then expiration, life and death. Every day really is a good day.[19]

Yet "anguish is everywhere," as the Buddha said. The source of that anguish is clinging. What is clinging? What is the object of clinging? It is cherishing the notion of permanence for this old fellow sitting here. It is cherishing the notion of independence from others. Clinging is the heart of my anguish and of human anguish. There is a release from this anguish, however, that Bokkei, for one, found for himself.

This release comes with practice, the Middle, or Eightfold, Path, from Right Views to Right Zazen. Right Zazen is, for example, counting the breaths—facing this point "one," this point "two," this point "three." The point of no magnitude is the marvelous void charged with brilliant light. The circular path leads round again from Right Zazen to Right Views, Right Thought, Right Speech, and the rest, to form our practice in daily life. On this Noble Path we find true human happiness that is far removed from the ordinary conspiracy of make-believe. It is a matter of finding ourselves temporarily all here together, resolving to take good care of each other—and doing it.

The Eightfold Path does not, however, directly address the grief that one experiences with letting go. My first inkling of the real nature of grief was something my Zen friend R. H. Blyth said to me one day: "I love my sister-in-law much more than I did when she was alive." I thought to myself, "How strange!"

Then many years later, Anne's mother died. We flew up to San Francisco for the funeral. She had lived in a large house in the Presidio district of San Francisco, all by herself after her husband died. The close relatives gathered on one side of the living room. The coffin was there on the other side. We stood around enjoying, as the occasion would allow, our reunion with each other. But there was a little too much chatter, so Anne and I went and knelt by the coffin and quietly recited the *Emmei Jikku Kannon Gyō*, the "Ten-Verse Kuan-yin Sūtra of Timeless Life." In those moments I felt her mother's presence far more intimately and vividly than I ever did when she was alive.

This is, it seems to me, the nature of grief. Beyond tears, beyond self-blame, it is the experience of the person, the presence of the person, and it is very poignant. For me, kneeling by her coffin, it was the pure grace of my mother-in-law.

One hears elderly people who have lost a spouse speak of "my angel husband" or "my saintly wife." I have tended to dismiss these allusions as sentimental, or even as denial. Now I understand their reality. After thirty-seven years of marriage, I knew Anne's inadequacies as she knew mine, as any family member knows the dark side of a sibling, parent, or spouse. But when she died, all her shortcomings abruptly vanished into thin air, just as her body vanished. Anne's shortcomings *were* her body, the physical barriers of her aspiration, just as my neuroses and foibles get in my own way.

I relate through flawed materials, and so does everyone else. With her death, however, Anne stands forth as the Nirmānakāya, the mysterious and joyous Buddha who is individually unique and pristine as herself. This is her gift, her dāna, which she entrusted to me and to her relatives and friends by dying. The bereaved old folks whose words I had dismissed as sticky sentiment and denial realized in their own ways a perennial mystery—the gift by death of the rare, singular person—and I know now that they spoke in genuine awe. Thus I take issue with Mark Antony when he declaims that the evil that we do lives after us, while the good often lies interred with our bones. It's surely the other way around.

Recently one of my middle-aged friends shared with me his unhappiness about his father. His mother had died a year before, and his father had fallen into depression. He then had a stroke and then another stroke. "Of course," my friend observed unexpectedly, "they loved each other, but they argued a lot." To quote Issa again, this time with sympathy:

> In the dewdrop
> of this dewdrop world,
> such quarrels![20]

When my friend linked his father's strokes to his mother's death and to the arguments his mother and father used to have, I heard the possibility that his father had at last experienced the true nature of his deceased wife and was consumed with regret that his realization had come too late. He fell into despair, and his decline began. One of my early Japanese friends, widowed for many years, said sadly to me, "I realize that I thought of my wife as a broom."

With one's own human failings, it is probably not possible to summon up full appreciation for the rare Buddha who is one's spouse or family member or friend. We can, however, practice abiding, patiently and lovingly, with the failings of the other, while acknowledging our own weaknesses and inadequacies.

I am resolved to learn from my bereavement and exercise loving patience more carefully with my other family members, sangha members, and friends—with the clerk at the post office, the cat, the dog, the hibiscus. These, too, I now know much more clearly, are also the Nirmānakāya. Each being is the Tathāgata, as the Buddha Shākyamuni said—the living Buddha who comes purely forth, a sister or a brother to protect and nourish.

1994

ETHICS AND REVOLUTION

The Path Beyond No-Self

Western Perspectives

Buddhist writings caution endlessly against concepts. The *Diamond Sūtra*, the inspiration of Zen students, is devoted to destroying all concepts, even the concept that concepts must be destroyed. In the *Diamond Sūtra* we find the Buddha advising his interlocutor Subuti and the rest of us not to be captivated by names and designated qualities like "Buddha Lands," "molecules," "galaxies," "enlightenment," "Tathāgata," and even such notions as "ego entity," "personality," and "exclusive individuality." The person who is not bound in such a way can be called one who abides joyously in peace.[1]

Moreover, while it is possible to be free of concepts—and this freedom can be joyous—the moment we dwell on the words "freedom" and "joy" we are trapped again. It is in no-freedom that we find freedom, in no-joy that we find joy. This is the Way of Zen or of freedom even from Zen.

Yet we must use words like "Zen," "self," "Buddha Lands" "galaxies," "molecules," in order to function socially. Naming is the primary human act. In the book of Genesis, Adam named the beasts of the field and the fowls of the air before he did anything else. Each infant creates the world in this way. The first two words my son strung together were "What's that?" Let me name that thing! This imperative has brought forth human civilization from the inchoate void.

The Buddha realized that the potential form of this inchoate void is harmony and conveyed his realization to us. Walking the dusty roads of the Ganges Valley, he turned the Dharma wheel of realization and accord right there in the untidy and unreliable context of coming and going, of dying and being born. As sons and daughters of the Buddha we turn that same wheel within our communities and in the home, the workplace, in social gatherings, and public meetings.

Just as the Buddha and his successors warn us about concepts from their Asian viewpoints, Western thinkers have also been concerned about the snare of words. Fritz Mauthner, a late-nineteenth-century philosopher and linguist, takes a position very close to that of the Buddha in the *Diamond Sūtra*. Mauthner points out that words cannot fully express reality; they are just categories into which sense impressions are sorted and stored and with which they are structured in the human memory.[2]

Human senses themselves are limited, Mauthner declares, and can only give us an anthropomorphic view of the world. Under different circumstances, our senses might have evolved differently, perhaps to include a better navigational sense—like that of birds, for example. We see that most birds have a navigational sense that is superior to ours. If we had such a sense, the structure of our memory would be different, and our vocabulary and grammar would express reality accordingly.

Our sorting and storing process can lead to further limitations and distortions, for our words take on a life of their own and influence each other, creating generalized ideas. We tend to "understand" new sense data by warping reality to fit these mental abstractions. (Since Mauthner's time, of course, scientists have come to understand how the observation is the observer and to regard their formulations not as law but as metaphor.)

Mauthner followed his line of argument to the ultimate position of "no-self," suggesting that our feeling of individuality is an illusion created by our unreliable senses and seemingly confirmed by the generalizations we can make from them. At the same time, he observed, this is the process that makes us human. Sense data and its language are the repositories of our tradition. The human self is a recurrence of tradition and a step in its continuation. When language is metaphorical and poetical, then it too is a renewal, though

Mauthner seems hesitant at this point, expressing only the hope that poetical language can present some valid contact with reality. The Zen master is far more positive:

> Yün-men addressed his students and said, "The old Buddhas and the pillar merge—on what level of mental activity is this?"
>
> Answering for them he said, "Clouds gather over North Mountain; rain falls on South Mountain."[3]

These are metaphors of one who has died completely to the world of abstractions—even such abstractions as time, form, space, and cause—and who seeks to present the vitality of such a death in a naked human encounter with the world. Mauthner could have shown Yün-men a thing or two about the nature of memory, but Yün-men was the master of words as the medium of all that is.

Fritz Mauthner was an important figure in the intellectual circles of fin-de-siècle Berlin, and he profoundly influenced Gustav Landauer, a younger contemporary who carried Mauthner's thought into realms that his biographer Charles B. Maurer calls "mystical anarchism." Both men understood the self to be simply a product of memory, including human racial memory, but Landauer had an inkling of the essential point of Zen Buddhism: that the whole world is a single psyche of countless differentiations and the individual self is an embodiment of all that is—the universe incarnate. It is not surprising that Landauer venerated Whitman and translated his work into German.[4]

Though the vision of both Landauer and Whitman contained multitudes, neither man could show how elements of the environment not merely *are* the self but *make* the self and that there is no other source of self. Compare Yün-men again:

> Yün-men addressed his disciples and said, "Each of you has your own light. If you try to see it, you cannot. The darkness is dark, dark. Now, what is your light?"
>
> Answering for them, he said, "The pantry, the gate."[5]

When you reflect upon yourself, you find nothing there at all. Try it. Take a moment. When you just breathe in and out quietly, what is your source? Your light? Really, there is nothing there. Then what is your source, your light? There it is! The pantry! There it is! The gate!

Our language is formed in turn by the pantry and the gate, by the rose and the lily, the dog and the fish and the tiger, the clouds and the stones. There is no language that is not the wording of our habitat, and no self whatever, except a temporary formation of the total environment. Nothing is absolute, and at this point Landauer seems closer to Hinduism than to Buddhism, postulating reality for the universal psyche, separate from its mortal qualities. For him, the psyche was something that does not die. This is the Senika heresy, named for a monk in the Buddha's time, and condemned by the Buddha in the *Mahāparinirvāna Sūtra* as a concept of an absolute that stands apart from the relative. The Senika heresy is also criticized in careful detail by Dōgen Zenji in his "Bendōwa."[6]

Nonetheless, Landauer's thought and indeed his life are profoundly instructive. "It is time," he wrote:

> for the insight that there is no individual, but only unities and communities.... Individuals are only manifestations and points of reference, electric sparks of something grand and whole.[7]

Landauer does not linger on sparks as metaphors for humanity but moves quickly to the flesh-and-blood reality of human beings as avatars of something grand and whole. This is where he is most instructive for Buddhists. He saw clearly the implications of selflessness, understood "consciousness" for what the word means— "consensual experience"—and knew "compassion" to be "suffering with others." Unlike our Buddhist ancestors, he was free to apply his metaphysics to the realms of sociology, politics, and economics. He wrote vividly of human suffering in words that might be written today, almost a hundred years later:

> The great mass of people is separated from the earth and its products, from the earth and the means of labor. People live in poverty and insecurity. They have no joy and meaning in their lives. They work in a way that makes them dull and joyless. Entire masses of people often have no roof over their head. They freeze, starve, and die miserably.[8]

It is *Geist* that is missing, Landauer says—communal spirit, the *volksseele*, or folk soul, the larger self of people in a particular region, culture, or nation. It dwells in the hearts of individuals who give themselves over to the unfolding of this spirit as they work together in communal units that interpenetrate to form a "society of societies." *Geist* can be compared with Plato's *philia*, the friendship of high-minded individuals who are drawn together by their affinity for noble conduct and their rejection of self-centered materialism.

For Plato, *philia* was a universal law that human beings cultivate for their fulfillment in society:

> Partnership and friendship, orderliness and self-control, and justice hold together heaven and earth, and that is why they call this universe a *world order*.[9]

For Landauer, *Geist* too had its ground in natural order, the totality of independent units found in the Middle Ages of Europe, the clan structure of traditional societies. Martin Buber, a disciple of Landauer, quotes his mentor and comments:

> "Such is the task of the socialists and of the movements they have started among the peoples: to loosen the hardening of hearts so that what lies buried may rise to the surface: so that what truly lives yet now seems dead may emerge and grow into the light." Men who are renewed in this way can renew society, and since they know from experience that there is an immemorial stock of community that has declared itself in them as something new, they will build into the new structure everything that is left of community-form.[10]

Deeper even than the clan structure of human racial memory lies the omnipresent tendency to unify, found throughout the universe in crystals and insects and stellar systems. In human terms, it is the way of organizing in small groups that cooperate in turn with other groups, coming into being, going out of being, encouraging the fulfillment of each individual avatar of the whole. Landauer looked back at the Middle Ages:

> A level of great culture is reached when manifold, exclusive, and independent communal organizations exist contempora-

neously, all impregnated with a uniform *Geist*, which does not reside in the organizations or arise from them, but which holds sway over them as an independent and self-evident force.[11]

This is the "society of societies," layered and contiguous in a network of agriculture, small industry, and cooperative finance, with support for and delight in art, music, and poetry. There was plenty of exploitation and suffering in the Middle Ages, of course, but there was also communal spirit, and Christianity was its medium.

Christianity, however, has lost its power. "Enough of that," Landauer says, "of those misunderstood remnants of a symbolism that once made sense."[12] Yet as Christianity loses its meaning, industrialism and mercantilism become our modern imperatives and *Geist* itself is suppressed, as Landauer observes:

> Even when [managers] know that the market can absorb their commodities only with difficulty or not at all, or at least not at the desired price, they must continue to bombard it with their products: for their production plants and enterprises are not guided by the needs of a coherent organic class of people of a community . . . , but by the demands of their production machinery, to which thousands of workers are harnessed like Ixion on the wheel.[13]

This intolerable situation is not inexorable fate, Landauer believed. With trust in one another, we can awaken our *Geist* and fulfill our potential as human beings. He organized the Socialist Bund Association in 1908, intended as the beginning of an alternate socioeconomic network, inspiring the formation of twenty small groups by 1911 in Berlin, Zurich, and other German and Swiss cities and even one in Paris. "We must return," he said,

> to rural living and to a unification of industry, craftsmanship, and agriculture, to save ourselves and learn justice and community. What Peter Kropotkin taught us about the methods of intensive soil cultivation and unification of intellectual and manual labor in his important and now famous book *The*

Field, the Factory, the Workshop, as well as the new form of credit and monetary cooperative, must all be tested now in our most drastic need and with creative pleasure.[14]

Like Landauer, we can find inspiration in the thought of the nineteenth-century anarchists—Kropotkin, Proudhon, and others—but it is Landauer, inspired by Mauthner, who understands most clearly that we should not literally "follow" the old teachers but use their words to remind ourselves of what we already know, evoking our long-frozen memory of community forms.

As Buddhists, our memory of community forms is embedded in the sangha, though the sangha has not always been organized in units as small as Landauer felt was ideal. It has not always been willing to form networks and has not always been free of worldly entanglements. Nonetheless, its aspirations in these respects have always been clear.

I propose that we form informal groups within our larger sanghas and examine what is going on in our world. The great multicentered self is being overwhelmed, at least on this Earth, by people who are impelled as individuals, corporations, and states to prove their personal authority and enhance their accumulation of wealth. We can bypass all this and empower ourselves in small, self-reliant groups of like-minded friends. We can work with socially relevant financial associations, worker-owned industries and farms, cooperative markets, little theaters, small presses and galleries—and we can create our own.

Our practice gives us unique readiness to realize, with Landauer, that while each of us is unique, we are not separate. We are organic elements of something far grander—and ready to put our understanding into practice. I trust, however, that we will not be working for some millennium of the future. As we labor together, using the most skillful and compassionate means we can find, our engagement itself will turn out to be our goal, the Way itself will turn out to be peace, and the milestones we reach, one after the next, will fulfill the Buddha's vows more and more fully and intimately.

Revised from a paper read at the conference "Toward an American Vinaya," Green Gulch Zen Center, Muir Beach, California, June 3–8, 1990.

Envisioning the Future

SMALL is beautiful," E. F. Schumacher said, but it was not merely size that concerned him. "Buddhist economics must be very different from the economics of modern materialism," he said. "The Buddhist sees the essence of civilization not in a multiplication of wants but in the purification of human character."[1]

Schumacher evokes the etymology of "civilization" as the process of civilizing, of becoming and making civil. Many neglect this ancient wisdom of words in their pursuit of acquisition and consumption, and those with some civility of mind find themselves caught in the dominant order by requirements of time and energy to feed their families. As the acquisitive system burgeons, its collapse is foreshadowed by epidemics, famine, war, and the despoliation of the earth and its forests, waters, and air.

I envision a growing crisis across the world as managers and their multinational systems continue to deplete finite human and natural resources. Great corporations, underwritten by equally great financial institutions, flush away the human habitat and the habitat of thousands of other species far more ruthlessly and on a far greater scale than the gold miners who once hosed down mountains in California. International consortia rule sovereign over all other political authority. Presidents and parliaments and the United Nations itself are delegated decision-making powers that simply carry out previously established agreements.

Citizens of goodwill everywhere despair of the political process.

The old enthusiasm to turn out on election day has drastically waned. In the United States, commonly fewer than 50 percent of those eligible cast a ballot. It has become clear that political parties are ineffectual—whether Republican or Democrat, Conservative or Labor—and that practical alternatives must be found.

We can begin our task of developing such alternatives by meeting in informal groups within our larger sanghas to examine politics and economics from a Buddhist perspective. It will be apparent that traditional teachings of interdependence bring into direct question the rationale of accumulating wealth and of governing by hierarchical authority. What, then, is to be done?

Something, certainly. Our practice of the Brahma Vihāras—kindliness, compassion, goodwill, and equanimity—would be meaningless if it excluded people, animals, and plants outside our formal sangha. Nothing in the teachings justifies us as a cult that ignores the world. We are not survivalists. On the contrary, it is clear that we're in it together with all beings.

The time has surely come when we must speak out as Buddhists, with firm views of harmony as the Tao. I suggest that it is also time for us to take ourselves in hand. We ourselves can establish and engage in the very policies and programs of social and ecological protection and respect that we have heretofore so futilely demanded from authorities. This would be engaged Buddhism, where the sangha is not merely parallel to the forms of conventional society and not merely metaphysical in its universality.

This greater sangha is, moreover, not merely Buddhist. It is possible to identify an eclectic religious evolution that is already under way, one to which we can lend our energies. It can be traced to the beginning of this century, when Tolstoy, Ruskin, Thoreau, and the New Testament fertilized the *Bhagavad Gita* and other Indian texts in the mind and life of M. K. Gandhi. The Southern Buddhist leaders A. T. Ariyaratne and Sulak Sivaraksa and their followers in Sri Lanka and Thailand have adapted Gandhi's "Independence for the Masses" to their own national needs and established programs of self-help and community self-reliance that offer regenerative cells of fulfilling life within their materialist societies.[2]

Mahayana has lagged behind these developments in South and

Southeast Asia. In the past, a few Far Eastern monks like Gyōgi Bo-satsu devoted themselves to good works, another few like Hakuin Zenji raised their voices to the lords of their provinces about the poverty of common people, and still others in Korea and China organized peasant rebellions, but today we do not see widespread movements in traditional Mahayana countries akin to the village self-help programs of Ariyaratne in Sri Lanka, or empowerment networks similar to those established by Sulak in Thailand.

"Self-help" is an inadequate translation of *swaraj*, the term Gandhi used to designate his program of personal and village independence. He was a great social thinker who identified profound human imperatives and natural social potentials. He discerned how significant changes arise from people themselves, rather than from efforts on the part of governments to fine-tune the system.

South Africa and Eastern Europe are two modern examples of change from the bottom up. Perceptions shift, the old notions cannot hold—and down come the state and its ideology. Similar changes are brewing, despite repressions, in Central America. In the United States, the economy appears to be holding up by force of habit and inertia in the face of unimaginable debt, while city governments break down and thousands of families sleep in makeshift shelters.

Not without protest. In the United States, the tireless voices of Ralph Nader, Noam Chomsky, Jerry Brown, and other cogent dissidents remind us and our legislators and judges that our so-called civilization is using up the world. Such spokespeople for conservation, social justice, and peace help to organize opposition to benighted powers and their policies and thus divert the most outrageous programs to less flagrant alternatives.

Like Ariyaratne and Sulak in their social contexts, we as Western Buddhists would also modify the activist role to reflect our culture as well as our spiritual heritage. But surely the Dharmic fundamentals would remain.[3] Right Action is part of the Eightfold Path that begins and ends with Right Meditation. Formal practice could also involve study, reciting the ancient texts together, Dharma discussion, religious festivals, and sharing for mutual support.

In our workaday lives, practice would be less formal and could include farming and protecting forests. In the United States, some of our leading intellectuals cultivate the ground. The distinguished poet W. S. Merwin has through his own labor created an arboretum of native Hawaiian plants at his home on Maui. He is thus restoring an important aspect of Hawaiian culture, in gentle opposition to the monocultures of pineapple, sugar, and macadamia nut trees around him. Another progressive intellectual, Wendell Berry, author of some thirty books of poetry, essays, and fiction, is also a small farmer. Still another reformative intellectual and prominent essayist, Wes Jackson, conducts a successful institute for small farmers. Networking is an important feature of Jackson's teaching. He follows the Amish adage that at least seven cooperating families must live near each other in order for their small individual farms to succeed.[4]

All such enterprise takes hard work and character practice. The two go together. Character, Schumacher says, "is formed primarily by a man's work. And work, properly conducted in conditions of human dignity and freedom, blesses ourselves and equally our products."[5] With dignity and freedom we can collaborate, labor together, on small farms and in cooperatives of all kinds—savings and loan societies, social agencies, clinics, galleries, theaters, markets, and schools—forming networks of decent and dignified modes of life alongside and even within the frames of conventional power. I visualize our humane network having more and more appeal as the power structure continues to fall apart.

This collaboration in networks of mutual aid would follow from our experience of *paticca-samuppāda*, interdependent co-arising. All beings arise in systems of biological affinity, whether or not they are even "alive" in a narrow sense. We are born in a world in which all things nurture us. As we mature in our understanding of the Dhamma, we take responsibility for paticca-samuppāda and continually divert our infantile expectations of being nurtured to an adult responsibility for nurturing others.

Buddhadāsa Bhikkhu says:

> The entire cosmos is a cooperative. The sun, the moon, and the stars live together as a cooperative. The same is true for humans and animals, trees and soil. Our bodily parts function as

a cooperative. When we realize that the world is a mutual, interdependent, cooperative enterprise, that human beings are all mutual friends in the process of birth, old age, suffering, and death, then we can build a noble, even heavenly environment. If our lives are not based in this truth, then we shall all perish.[6]

Returning to this original track is the path of individuation that transforms childish self-centeredness to mature views and conduct. With careful, constant discipline on the Eightfold Noble Path of the Dharma, greed becomes dāna, exploitation becomes networking. The root-brain of the newborn becomes the compassionate, religious mind of the elder. Outwardly the elder does not differ from other members; her or his needs for food, clothing, shelter, medicine, sleep, and affection are the same as anyone else's. But the elder's smile is startlingly generous.

It is a smile that rises from the Buddha's own experience. Paticcasamuppāda is not just a theory but the profound realization that I arise with all beings and all beings arise with me. I suffer with all beings; all beings suffer with me. The path to this fulfillment is long and sometimes hard; it involves restraint and disengagement from ordinary concerns. It is a path that advances over plateaus on its way, and it is important not to camp too long on any one plateau. That plateau is not yet your true home.

Dharmic society begins and prevails with individuals walking this path of compassionate understanding, discerning the noble option at each moment and allowing the other options to drop away. It is a society that looks familiar, with cash registers and merchandise, firefighters and police, theaters and festivals, but the inner flavor is completely different. Like a Chinese restaurant in Madras: the decor is familiar, but the curry is surprising.

In the United States of America, the notion of compassion as the touchstone of conduct and livelihood is discouraged by the culture. Yet here and there one can find Catholic Workers feeding the poor, religious builders creating housing for the homeless, traditional people returning to their old ways of agriculture.

Small is the watchword. Huge is ugly, as James Hillman has

pointed out.[7] Huge welfare goes awry, huge housing projects become slums worse than the ones they replace, huge environmental organizations compromise their own principles in order to survive, huge sovereignty movements fall apart with internal dissension. The point is that huge *anything* collapses, including governments, banks, multinational corporations, and the global economy itself—because all things collapse. Small can be fluid, ready to change.

The problem is that the huge might not collapse until it brings everything else down with it. Time may not be on the side of the small. Our awareness of this unprecedented danger impels us to take stock and do what we can with our vision of a Dharmic society.

The traditional sangha serves as a model for enterprise in this vision. A like-minded group of five can be a sangha. It can grow to a modest size, split into autonomous groups, and then network. As autonomous lay Buddhist associations, these little communities will not be sanghas in the traditional sense but will be inheritors of the name and of many of the original intentions. They will also be inheritors of the Base Community movements in Latin America and the Philippines—Catholic networks that are inspired by traditional religion and also by nineteenth-century anarchism.[8] Catholic Base Communities serve primarily as worship groups, study groups, moral support societies, and nuclei for social action. They can also form the staff and support structure of small enterprises.

The Catholic Base Community is grounded in Bible study and discussions. In these meetings, one realizes for oneself that God is an ally of those who would liberate the poor and oppressed. This is liberation theology of the heart and gut. It is an internal transformation that releases one's power to labor intimately with others to do God's work.[9]

The Buddhist counterpart of Bible study would be the contemplation and realization of paticca-samuppāda, of the unity of such intellectual opposites as the one and the many found in Zen practice, and the interdependence presented in the sacred texts, such as the *Hua-yen ching*.[10] Without a literal God as an ally, one is thrown back on one's own resources to find the original track, and there one finds the ever-shifting universe with its recurrent metaphors of interbeing to be the constant ally.

There are other lessons from liberation theology. We learn that we need not quit our jobs to form autonomous lay sanghas. Most Base Communities in Latin America and the Philippines are simply groups that have weekly meetings. In Buddhist countries, coworkers in the same institution can come together for mutual aid and religious practice. In the largest American corporations, such as IBM, there will surely be a number of Buddhists who could form similar groups. Or we can organize co-housing arrangements that provide for the sharing of home maintenance, child care, and transportation and thus free up individuals for their turns at meditation, study, and social action. Buddhist Peace Fellowship chapters might consider how the Base Community design and ideal could help to define and enhance their purposes and programs.[11]

Thus it wouldn't be necessary for the people who work in corporations or government agencies to resign when they start to meet in Buddhist Base Communities. They can remain within their corporation or government agency and encourage the evolution and networking of communities, not necessarily Buddhist, among other corporations and agencies. Of course, the future is obscure, but I find myself relating to the mythology of the Industrial Workers of the World—that as the old forms collapse, the new networks can flourish.

Of course, the collapse, if any, is not going to happen tomorrow. We must not underestimate the staying power of capitalism. Moreover, the complex, dynamic process of networking cannot be put abruptly into place. In studying Mondragón, the prototype of large, dynamic cooperative enterprise in the three Basque counties of northern Spain, William and Kathleen Whyte counted more than a hundred worker cooperatives and supporting organizations with 19,500 workers in 1988. These are small—even tiny—enterprises, linked by very little more than simple goodwill and a profound sense of the common good. Together they form a vast complex of banking, industry, and education that evolved slowly, if steadily, from a single class for technical training set up in 1943.[12]

We must begin with our own training classes. Mondragón is worth our study, as are the worker-owned industries closer to home—for example, the plywood companies in the Pacific Northwest. In 1972 Carl Bellas studied twenty-one such companies whose inner structures consisted of motivated committees devoted to the many aspects of production and whose managers were responsible to a general assembly.[13]

In the course of our training classes, it is also essential that we examine the mechanism of the dominant economy. Usury and its engines have built our civilization. The word "usury" has an old meaning and a modern one. In the spirit of the old meaning of usury—lending money at interest—the banks of the world, large and small, have provided a way for masses of people for many generations across the world to own homes and to operate farms and businesses. In the spirit of the modern meaning of usury, however—the lending of money at *excessive* interest—a number of these banks have become gigantic, ultimately enabling corporations almost as huge to squeeze small farmers from their lands, small shopkeepers from their stores, and to burden homeowners with car and appliance payments and lifetime mortgages.

For over 1,800 years, the Catholic church had a clear and consistent doctrine on the sin of usury in the old sense of simply lending money at interest. Nearly thirty official church documents were published over the centuries to condemn it.

Out of the other side of the Vatican, however, came an unspoken tolerance for usury so long as it was practiced by Jews. The church blossomed as the Medici family of bankers underwrote the Renaissance, but at the same time, pogroms were all but sanctioned. The moral integrity of the church was compromised. Finally, early in the nineteenth century, this kind of hypocrisy was abandoned—too late in some ways, for the seeds of the Holocaust had already been planted. Today the pope apologizes to the Jews, and even the Vatican has its bank.[14] Usury in both old and modern implications is standard operating procedure in contemporary world culture.

Like the Medicis, however, modern bankers can be philanthropic. In almost every city in the United States, bankers and their institutions are active in support of museums, symphony orchestras, clinics, and schools. Banks have almost the same social function as traditional Asian temples: looking after the poor and promoting cultural activities. This is genuine beneficence, and it is also very good public relations.

In the subdivisions of some American cities, such as the Westwood suburb of Los Angeles, the banks even look like temples. They are indeed the temples of our socioeconomic system. The banker's manner is friendly yet his interest in us is, on the bottom line, limited to the interest he extracts from us.

One of the banks in Hawai'i has the motto "We say 'Yes' to you," meaning "We are eager for your money." Their motto is sung interminably on the radio and TV, and when it appears in newspapers and magazines we find ourselves humming the tune. Similar lightweight yet insidious persuasions are used with Third World governments for the construction of freeways and hydroelectric dams and administrative skyscrapers.

Governments and developers in the Third World are, in fact, the dupes of the World Bank and the International Monetary Fund (IMF):

> It is important to note that IMF programs are not designed to increase the welfare of the population. They are designed to bring the external payments account into balance. . . . The IMF is the ultimate guardian of the interests of capitalists and bankers doing international business.[15]

These are observations of the economist Kari Polyani Levitt, quoted as the epigraph of a study entitled *Banking on Poverty*. The editor of this work concludes that policies of the IMF and the World Bank "make severe intrusions upon the sovereign responsibilities of many governments of the Third World. These policies not only often entail major additional cuts in the living standards of the poorest sectors of Third World societies but are also unlikely to produce the economic results claimed on their behalf."[16]

Grand apartment buildings along the Bay of Bombay show that the First World with its wealth and leisure is alive and well among the prosperous classes of the old Third World. The Third World with its poverty and disease flares up in cities and farms of the old First World. In *The Prosperous Few and the Restless Many*, Noam Chomsky writes:

> In 1971, Nixon dismantled the Bretton Woods system, thereby deregulating currencies.[17] That, and a number of other changes, tremendously expanded the amount of unregulated capital in the world and accelerated what's called the globalization of the economy.
>
> That's a fancy way of saying that you can export jobs to high-repression, low-wage areas.[18]

Factories in South Central Los Angeles moved to Eastern Europe, Mexico, and Indonesia, attracting workers from farms. Meantime, victims in South Central Los Angeles and other depressed areas of the United States, including desolate rural towns, turn in large numbers to crime and drugs to relieve their seemingly hopeless poverty. One million American citizens are currently in prison, with another two million or so on parole or probation. More than half of these have been convicted of drug-related offenses.[19] It's going to get worse. Just as the citizens of Germany elected Hitler chancellor in 1932, opening the door to fascism quite voluntarily, so the citizens of the United States have elected a Congress that seems bent on creating a permanent underclass, with prison expansion to provide much of its housing.

Is there no hope? If big banks, multinational corporations, and cooperating governments maintain their strategy to keep the few prosperous and the many in poverty, then where can small farmers and shopkeepers and managers of clinics and social agencies turn for the money they need to start up their enterprises and to meet emergencies? In the United States, government aid to small businesses and farms, like grants to clinics and social agencies, is being cut back. Such aid is meager or nonexistent in other parts of the world, with notable exceptions in northern Europe.

Revolving credit associations called *hui* in China, *kye* in Korea, and *tanamoshi* in Japan have for generations down to the present provided start-up money for farmers and owners of small businesses, as well as short-term loans for weddings, funerals, and tuition. In

Siam there are rice banks and buffalo banks designed for sharing resources and production among the working poor.[20] The Grameen banks of Bangladesh are established for the poor by the poor. Shares are very tiny amounts, amounting to the equivalent of just a few dollars, but in quantity they are adequate for loans at very low interest to farmers and shopkeepers.[21]

Similar traditional cooperatives exist in most other cultures. Such associations are made up of like-minded relatives, friends, neighbors, coworkers, or alumnae. Arrangements for borrowing and repayment among these associations differ, even within the particular cultures.[22] In the United States, cooperatives have been set up outside the system, using scrip and labor credits—most notably, Ithaca Hours, involving 1,200 enterprises. The basic currency in the latter arrangement is equal to ten dollars, considered to be the hourly wage. It is guaranteed by the promise of work by members of the system.[23]

We can utilize such models and develop our own projects to fit our particular requirements and circumstances. We can stand on our own feet and help one another in systems that are designed to serve the many, rather than to aggrandize the wealth of the few.

Again, small is beautiful. Whereas large can be beautiful too, if it is a network of autonomous units, monolithic structures are problematic even when fueled by religious idealism. Islamic economists theorize about a national banking system that functions by investment rather than by a system of interest. However, they point out that such a structure can only work in a country where laws forbid lending at interest and where administrators follow up violations with prosecution.[24] So for those of us who do not dwell in certain Islamic countries that seek to take the Koran literally, such as Pakistan and some of the Gulf states, the macrocosmic concept of interest-free banking is probably not practical.

Of course, revolving credit associations have problems, as do all societies of human beings. There are defaults, but peer pressure among friends and relatives keep these to a minimum. The discipline of Dharma practice would further minimize such problems in a Buddhist loan society. The meetings could be structured with ritual and Dharma talks to remind the members that they are practicing the virtues of the Buddha Dharma and bringing paticca-samuppāda into play in their workaday lives. They are practicing trust, for all beings are the Buddha, as Hakuin Zenji and countless other teachers remind us.[25] Surely only serious emergencies would occasion a delinquency, and contingency planning could allow for such situations.

Dharma practice could also play a role in the small Buddhist farm or business enterprise. In the 1970s, under the influence of Buddhists, the Honest Business movement arose in San Francisco. This was a network of small shops whose proprietors and assistants met from time to time to encourage one another. Their policy was to serve the public and to accept enough in return from their sales to support themselves, sustain their enterprises, and pay the rent. Their account books were on the sales counters, open to their customers.[26]

The movement itself did not survive, though progressive businesses here and there continue the practice of opening their account books to customers.[27] Apparently the Honest Business network was not well enough established to endure the change in culture from the New Age of the 1970s to the pervasive greed of the 1980s. I suspect there was not a critical mass in the total number of shops involved, and many of them might have been only marginal in their commercial appeal. Perhaps religious commitment was not particularly well rooted. Perhaps also there was not the urgency for alternatives that might be felt in the Third World—an urgency that will surely be felt in all worlds as the dominant system continues to use up natural resources.[28] In any case, we can probably learn from the Honest Business movement and avoid its mistakes.

In establishing small enterprises—including clinics and social agencies and their networks—it is again important not to be content with a plateau. The ordinary entrepreneur, motivated by the need to support a family and plan for tuition and retirement, scrutinizes every option and searches out every niche for possible gain. The manager of an Honest Business must be equally diligent, albeit motivated by service to the community as well as by the family's needs.

Those organizing to lobby for political and economic reforms must also be diligent in following through. The Base Communities throughout the archipelago that forms the Philippines brought down the despot Ferdinand Marcos, but the new society wasn't ready to fly and was put down at once. The plateau was not the peak, and euphoria gave way to feelings of betrayal. However, you can be sure that many of those little communities are still intact. Their members have learned from their immediate history and continue to struggle for justice.

A. J. Muste, the great Quaker organizer of the mid twentieth century, is said to have remarked, "There is no way to peace; peace is the way." For our purposes, I would reword his pronouncement: "There is no way to a just society; our just societies are the way." Moreover, there is no plateau to rest on, only the inner rest we feel in our work and in our formal practice.

This inner rest is so important. In the short history of the United States, there are many accounts of utopian societies. Almost all of them are gone—some of them lasted only a few weeks. Looking closely, I think we can find that many of them fell apart because they were never firmly established as religious communities. They were content to organize before they were truly organized.

Families fall apart almost as readily as intentional communities these days, and Dharma practice can play a role in the household as well as in the sangha. As Sulak Sivaraksa has said, "When even one member of the household meditates, the entire family benefits."[29] Competition is channeled into the development of talents and skills; greed is channeled into the satisfaction of fulfillment in work. New things and new technology are used appropriately and are not allowed to divert time and energy from the path of individuation and compassion.

New things and new technology are very seductive. When I was a little boy, I lived for a time with my grandparents. These were the days before refrigerators, and we were too far from the city to obtain ice. So under an oak tree outside the kitchen door we had a cooler—a kind of cupboard made mostly of screen, covered with burlap that trailed into a pan of water. The burlap soaked up the water, and evaporation kept the contents of the cupboard cool, the milk fresh, and the butter firm. We didn't need a refrigerator. I can only assume that the reason my grandparents ultimately purchased one in later years was because they were persuaded by advertisements and by their friends.

We too can have coolers just outside the kitchen door or on the apartment veranda, saving the money the refrigerator would cost to help pay for the education of our children. Like our ancestors, we too can walk or take public transportation. We can come together like the Amish and build houses for one another. We can join with our friends and offer rites of passage to sons and daughters in their phase of experimenting and testing the limits of convention.

Our ancestors planned for their descendants; otherwise we might not be here. Our small lay Buddhist societies can provide a structure for Dharma practice, as well as precedent and flexible structures for our descendants to practice the Dharma in turn, for the next ten thousand years.

In formally sustaining the Dharma, we can also practice sustainable agriculture, sustainable tree farming, sustainable enterprise of all kinds. Our ancestors sustain us; we sustain our descendants. Our family members and fellow workers nurture us, and we nurture them—even as dāna was circulated in ancient times.

Circulating the gift, the Buddhist monk traditionally offers the Dharma, as we offer him food, clothing, shelter, and medicine. But he also is a bachelor. Most of us cannot be itinerant mendicants. Yet as one who has left home, the monk challenges us to leave home as well—without leaving home. There are two meanings of "home" here. One could be the home of the family, but with the distractions that obscure the Dharma. The other may involve the family but is also the inner place of peace and rest, where devotion to the Buddha Way of selflessness and affection is paramount. The monks and their system of dāna are, in fact, excellent metaphorical models for us. The gift is circulated, enhancing character and dignity with each round. Festivals to celebrate the rounds bring joy to the children and satisfaction to the elders.

I don't suggest that the practice of circulating the gift will be all sweetness and light. The practice would also involve dealing with mean-spirited imperatives, in oneself and in others. The Buddha and his elder leaders made entries in their code of vinaya (moral teachings) after instances of conduct that were viewed as inappropriate. Whether the Buddhist Base Community is simply a gathering of like-minded followers of the Dharma that meets for mutual support and study, whether it has organized to lobby for justice, or whether it conducts a business, manages a small farm, or operates a clinic, the guidelines must be clear. General agreements about what constitutes generous conduct and procedure will be valuable as references. Then, as seems appropriate, compassionate kinds of censure for departing from those standards could gradually be set into place. Guidelines should be set for conducting meetings, for carrying out the work, and for networking. There must be teaching, ritual, and sharing. All this comes with trial and error, with precedent as a guide but not a dictator.

Goodwill and perseverance can prevail. The rounds of circulating the gift are as long as ten thousand years, as brief as a moment. Each meeting of the little sangha can be a renewal of practice, each workday a renewal of practice, each encounter, each thought-flash. At each step of the way we remember that people and indeed the many beings of the world are more important than goods.

Revised from a paper read at the conference "Dhammic Society: Toward an INEB Vision," International Network of Engaged Buddhists, Wongsanit Ashram, Ongkharak Nakhom Nayok, Thailand, February 20–24, 1995.

The Experience of Emptiness

Use and Misuse

ZEN Buddhism is by definition indefinable, and in the context of its study, nothing—not even nothing—can be defined. In the *Diamond Sūtra*, we are told that there is no formulation of consummate truth. The Buddha himself, herself, or itself cannot be distinguished by any characteristic whatever.[1] Huang-po says:

> This spiritually enlightening nature is without beginning, as ancient as the Void, subject neither to birth nor to destruction, neither existing nor not existing, neither impure nor pure, neither clamorous nor silent, neither old nor young, occupying no space, having neither inside nor outside, size nor form, color nor sound. It cannot be looked for or sought, comprehended by wisdom or knowledge, explained in words, contacted materially or reached by meritorious achievement. All the Buddhas and Bodhisattvas, with all wriggling things possessed of life, share in this great Nirvanic nature.[2]

Despite such vivid cautions, some students understand this empty nature conceptually, and risk getting stuck in an undifferentiated place where correct and incorrect are the same, where male and female are the same—where all configurations disappear into a kind of pudding. The great teachers of the past addressed this risk directly:

The venerable Yen-yang asked Chao-chou, "When one has brought not a single thing, what then?"
Chao-chou said, "Put it down."[3]

When you cling to nothing as something, then you yourself are not truly empty, and the emptiness you cherish is no more than an idea. With this *notion* of emptiness, you can be persuaded that the homeless are an illusion, the rain forests are not being destroyed, there are no traditional peoples who are dying out, there is no one freezing or starving or dying from shrapnel in the former Yugoslavia. When you run over a child with your car, there is no child, after all. Put down that "not a single thing" or your successors will use it to enhance and support brutality and imperialism.

Indeed, for some of our ancestors and contemporaries, the mental discipline of Zen can be divorced from the compassion and wisdom of Buddhism. The Dharma becomes like a potted plant. So long as it is not hindered, it can be moved around and allowed to flower and bear fruit anywhere. The eminent scholar D. T. Suzuki is open to criticism on this point, where he writes:

> Zen has no special doctrine or philosophy, no set of concepts or intellectual formulas, except that it tries to release one from the bondage of birth and death by means of certain intuitive modes of understanding peculiar to itself. It is, therefore, extremely flexible in adapting itself to almost any philosophy and social doctrine as long as its intuitive teaching is not interfered with. It may be found wedded to anarchism or fascism, communism or democracy, atheism, or idealism, or any political or economic dogmatism. It is, however, generally animated with a certain revolutionary spirit, and when things come to a deadlock—as they do when we are overloaded with conventionalism, formalism, and other cognate isms—Zen asserts itself and proves to be a destructive force. The spirit of the Kamakura era was in this respect in harmony with the virile spirit of Zen.[4]

Is the social responsibility of Zen limited to the destruction of convention? Of course, it is important not to get locked into formal propriety, but is *breaking out* our only function? Is there really something to be called Zen that can accommodate itself to fascism? Surely there are conventional social standards and political and economic structures that are in keeping with the Buddha's vision of harmony. With all respect to my dear old Sensei, I would step down from my podium, abandon my heritage and my sangha, and wander as a root-less pilgrim if I agreed with the perspective of Zen Buddhism that he sets forth here.

Compare Dr. Suzuki's words with those of the Dalai Lama, who understands very well how all forms are empty of substance and how at the same time they come forth, precious in themselves. In his pub-lic talks he declares again and again, like Tōrei Zenji, that even our so-called enemies can be our teachers.

This is the true teaching of the Buddha. All beings come forth sa-cred in their suchness, and it is my responsibility and yours to make this clear. Adversaries, enemies—as metaphors and as folks out there in opposition to us—are not merely empty. They are the Buddha. When I experience myself as an empty Buddha, I am large, a bound-less container of multitudes. The challenge is to forget myself, to let my body and mind drop away and to encourage the body and mind of others to drop away, and to continue this dropping away endlessly, as Dōgen Zenji has said.[5] Then my practice of including more and more others will be endless too.

Dr. Suzuki was not the first to declare Zen to be something that can be used for authoritarian ends. The Kamakura shogunate was very impressed by the Zen teachers who had taken refuge in Japan from the turmoil in China at the end of the Sung dynasty, and en-couraged these teachers and their immediate Japanese successors to adapt the teachings to needs of the samurai. Addressing one of these samurai in his sangha, Takuan Sōhō Zenji used Zen terminology to sanitize bloodshed:

> The uplifted sword has no will of its own, it is all of empti-ness. . . . The man who is about to be struck down is also of emptiness, as is the one who wields the sword.[6]

Emptiness indeed! What about the blood? What about the wails of the widow and her children? Empty too, I suppose. It must be said that this is Buddhist antinomianism.

To be fair to Takuan Zenji, he offered his advice within the context of Japanese Buddhism, which was founded by Shōtoku Taishi in the seventh century as a means of maintaining the political structure of the country.[7] Moreover, he was not the first to declare deadly violence to be empty and therefore acceptable. In the Hindu scriptures, we find Krishna advising Arjuna that there is no killer and no one to be killed.[8] We also find such ideas flowering in the words and deeds of samurai after Takuan's time—for example, in those of Miyamoto Musashi and Yamaoka Tesshū, and today in the teachings and practice found in the International Zen Dōjō movement in Japan and in the West.[9]

It is our task, it seems to me, to take up Buddhism as a religion of infinite compassion, which Dr. Suzuki at his best realized very clearly.[10] As an accomplished Zen student said to me recently, "The experience of profound emptiness is at once the experience of great compassion." That's right, and the samurai Zen people are wrong! Do you say there is no right or wrong? Wrong!

On the traditional Zen Buddhist altar, Shākyamuni occupies the center seat; Mañjushrī, the incarnation of great wisdom, sits on one side; Samantabhadra, the incarnation of great action, sits on the other. These archetypes and their positions can be profoundly meaningful for the Buddhist pilgrim.

The great wisdom of Mañjushrī is the realization that everything is totally empty, vacant, void. Nothing abides: not the body, not the self, not the soul. With this realization one finds vast and joyous liberation, as we learn in the Four Noble Truths.

In her great action, Samantabhadra wields skillful means of demeanor, words, and deeds to turn the Dharma wheel with all beings. We devote ourselves to uncovering her talent as our own and to encouraging others to liberate themselves.

The position of these images on the Zen Buddhist altar recalls the Buddha Shākyamuni beneath the Bodhi tree, realizing the true nature of all things and arising to seek out his old friends. My teacher, Yasutani Haku'un Rōshi, used to begin his orientation to Zen practice with the Buddha's great realization: "Wonderful, wonderful! Now I see that all beings are the Tathāgata"—that is, all beings come forth as the Buddha himself, herself, itself. "Only their delusions and preoccupations keep them from testifying to that fact."[11]

If all beings are inherently Buddha and if we follow the way of Shākyamuni and encourage others to realize and testify to this fact, then surely the next step is to take positions of compassionate resistance to fascism and other repressive systems and to search out alternate structures for the application of the Buddha's experience—of our experience.

The Buddha's teachings include the Brahma Vihāras, the Noble Abodes of loving-kindness, compassion, joy in the liberation of others, and equanimity or impartiality. These four modes of selfless practice form part of the Dhyāna Pāramitā, the Perfection of Settled, Focused Meditation. Monks and nuns in ancient times absorbed themselves in these four ideals as practice on their cushions.[12]

As Buddhism evolves and our understanding matures, however, I think we can see that dhyāna is not just temple practice. Everything in the world is in dhyāna. Even people in a mental hospital are absorbed—not on a productive track, maybe, but still they are immersed in their crazy ways. The dog is absorbed in being a dog, the stone in being a stone.

We can also see that loving-kindness and the other Brahma Vihāras are more than just meditation themes. When we devote ourselves to the Buddha Way, we practice loving-kindness, compassion, sympathetic joy, and equanimity in the market and in our households. These ancient ideals inspire us as we write to our friends and as we greet Mormons when they come to the door to missionize. We internalize these ideals and extend tender care, as Tōrei Zenji advises us, to beasts and birds—and indeed to plants, pancakes, orange juice, and undershirt.[13] The haiku poet Issa wrote:

> Don't kill it!
> The fly wrings its hands;
> It wrings its feet.[14]

Samantabhadra, busy turning her wheel of wisdom and compassion, would be disappointed with the folks who veer off into philosophical nihilism or political expedience. I believe she would want to step beyond even Tōrei Zenji and Kobayashi Issa in our modern times, for we as citizens of the world face serious, socially systemic problems.

If we are to find peace in emptiness, if we are to take responsibility for our lives, then what is to be done? Great action, certainly, to the limits of our talent and beyond. But what action? It seems to me that the way of the future for Buddhists and their friends lies at three levels of reempowerment: the personal, the communal, and the global.

First, at the personal level, you and I must put aside any dependence we might have on others to do our religious work for us and take responsibility for the practice of emptying ourselves—so that we may fill ourselves with all beings. Moreover we can encourage this ancient, authentic practice among others. This is rigorous, exacting work and requires the guidance of someone who has traveled the path before.

I wrote once that the Buddha was a great autodidact, but really he wasn't. He lived with the best teachers of his time before he experimented with asceticism and meditation. Then when he could stand on his own feet, he did not speak from a vacuum. The wisdom of his old teachers, however transformed and transmuted, informed his words, we can be sure.

The good teacher is necessary for two reasons. She or he will encourage you and offer you guidance. She or he will also deny you the complacency of a plateau and urge you on to the peak of your potential and even beyond. Too often I meet people who have the confidence that comes with a spontaneous spiritual experience outside any discipline or practice. When I check them and tell them, "Not enough," they tend to become angry and to argue. Sometimes they disappear, which is too bad. So faith in the teacher is important. If she is worth her salt, she *knows*, and you must swallow hard and accept the fact that you probably don't have it yet.

The human tendency is to be satisfied with a milestone. But there are milestones after milestones without end. One of the great contributions of Zen Buddhism to world religion can be summed up with the words "Not yet, not enough, not yet enough!"

Second, at the communal level we can turn to the Buddhist sangha for our reempowerment. It is clear that the Buddha considered the sangha to be the only possible mode of universal realization. In his day, and in his tradition down to modern times, the lay community has supported the sangha of ordained monks and thus encouraged the growth and spread of the Dharma. But here we appear as Mahayana students in modern times. The lay community *is* the sangha, including spouses and children. We can encourage each other as parents, sangha members, and citizens to stand together and resist the benighted forces that are destroying cultures and ruining the planet.

Finally, at the global level of reempowerment, we can take responsibility as citizens of the world by finding within ourselves the seeds of other religions and cultures—then nurturing those seeds and encouraging them to grow and bear fruit. The mullahs and the patriarchs in the former Yugoslavia do not generally speak to each other, and this exclusive kind of silence is surely one cause of the terrible civil war that rages there, on and on. With interreligious encounter and dialogue, and with interethnic and intercultural encounter and dialogue—and with follow-up by mail, telephone, fax, and modem—we will surely find ways to create peace together across formerly impenetrable frontiers.

Compassion is our guide. The mental discipline of Zen is really spiritual and cannot be divorced from its Buddhist roots. We must see clearly when others attempt to use the experience of emptiness inconsistently with the Buddha's teachings. At such times, we must remember our heritage and the rationale of discipline. Back to A. J. Muste: "Peace is the Way!"

Revised from a paper read at the conference "Buddhists and Christians for Justice, Peace, and the Integrity of the Earth," Lassalle Haus, Bad Shönbrunn, Edilbach/Zug, Switzerland, July 17–22, 1994.

Brahmadanda, Intervention,
and Related Considerations

A Think Piece[1]

In July 1964, our resident monk departed under a cloud from the Koko An Zendō, leaving two women students in the mental health ward of the Queen's Medical Center. In the ensuing thirty and more years, I have been musing—and occasionally speaking out—on the subject of sexual exploitation of students by Buddhist teachers, searching for an appropriate role for myself in confronting it. The subject is difficult because, as John Bradford remarked as he watched criminals being led to the executioner, "There but for the grace of God go I."[2]

Occasionally a woman comes to the interview room wearing a particularly low-cut dress, so that when she bows before me, she might partially expose her breasts. In the early days, I would shut my eyes for the crucial moment, then open them again before she made eye contact. A fellow teacher said, "Why don't you keep your eyes open? Are you so susceptible?"

"No," I thought to myself, "surely I'm immune by now." So I tried keeping my eyes open and found that I was indeed vulnerable after all. I noticed that the sexual charge I got from that glimpse of pretty breasts would color my attitude and give the interaction an undesirable personal tug. So I returned to my old custom of closing my eyes.

I'm seventy-eight years old now, the fires are banked, but incidents almost every day remind me that under the ashes the coals are still glowing. I hug students by way of greeting them at potluck suppers and other informal occasions at our zendō. Do I hug the men the way I hug the women? Do I hug the old ladies the way I do their daughters or their nubile granddaughters? I can't be sure, but I do practice uniformity in hugging as best I can. Sometimes my best is none too good, and I exchange a rueful glance with my student. No need for words. The message is clear: "Sure, it's there, and that's where we'll leave it!"

The late-twelfth-century Lin-chi master Sung-yüan asked, "Why is it that someone of great satori does not cut off the vermilion thread?"[3] Vermilion was a color associated with women's undergarments in old China and thus with sexual energy. Sung-yüan's question is intimate, and as a kōan it requires an intimate response. As a comment, I would add that it is possible to cut the thread, but if I take such a drastic step, or even try to take it, then I'm dead, whether or not I am still walking around. That sexy vermilion, those hot encapsulations of vitality that glow under the ashes of decades, nurture and enrich my body, speech, and thought. They are my id, my alligator mind, my passion. Keep the home fires burning!

It would seem that the problem is not with the fire itself but with the fireplace, the container, the character. The welder's torch regulates its own fire and builds the human habitat, but the conflagration that destroys a city building has run wild. Lives and careers are ravaged.

We build character to contain the fire in our practice. Then the fire empowers body, speech, and thought. With the container in place, human passion saves the many beings.

One thing leads to another, as my grandmother used to say. Natural, healthy sexual attraction can lead to courtship in other circumstances, but for the Buddhist teacher working with students, I am sure that the practice must be to "leave it alone." Let it glow in peace, transmuting into bows, smiles, and words that encourage and inspire. Otherwise, the attraction becomes a grotesque mirror of courtship, leading step by step to tragedy.

It is possible to distinguish the case of a teacher who falls in love with a student and has an affair with her from that of a teacher with a record of seductions that reflects an obsession with sexual domination. However, many questions remain. Will the consent of the student be meaningful? Or will she simply have allowed herself to be persuaded by her transference to the teacher?

The honorable therapist who falls in love with his client will determine first if an affair would bring harm and then will terminate the therapy before entering into a love relationship. Termination is not an option for the Buddhist teacher, however, for in effect it involves expelling the student from the sangha. This has happened more than once, but the ensuing uproars have created enduring harm.

I am convinced (though a well-known marriage between a Zen Buddhist teacher and his student would seem an exception to prove the rule) that unless the teacher is ready to resign, he should be strict with himself and rigorously avoid the little preliminary steps that could lead to an affair, however deep the feelings involved might be. No private tea, no walk in the park. The bottom line is the health and practice of the student and of the Buddha sangha. The dynamics of transference can create havoc when they are disturbed.

On the other hand, I can visualize that as a one-time incident, the disruption created by a love affair between the Buddhist teacher and student could be taken up in reconciliation programs, with everyone learning and maturing in the process. It could lead to guidelines that set forth ways to avoid such exigencies. I don't seek to minimize the distress the love affair could cause, but it would, I think, be more moderate on the scale of human suffering than the widespread anguish created by willful actions that stand in for love but that are actually ruthlessly exploitative.

These problems are by no means unique to the Buddha sangha; sexual harassment is far-reaching in North America, and indeed across the world. Talk to any school counselor, any pediatrician, any social worker or psychologist. The same dynamics are at work in the home, academic seminar, science lab, doctor's office, or sanctuary. Moreover, the depressed position of women in the merging streams of Asian and European cultural history forms a precedent that contributes to sexual abuse in the Buddhist temple. However, the particular kind of transference that the unscrupulous Buddhist teacher exploits is especially anguishing for his victims.

Fortunately, victims now have places to turn. Networks have been set up by survivors of sexual abuse who offer support and suggestions. A prime resource is Survivor Connections, which publishes the journal *The Survivor Activist*, in which sexual abuse survivors tell their stories. Many articles tell of abuse by religious leaders, and over a hundred agencies and collectives devoted to the problems of sexual abuse are listed with addresses.[4]

The Buddhist community as a whole and Buddhist teachers especially have a particular role to play in dealing with the problem. As teachers or senior students, we can use our positions to effect change within the community and bring to the resolution process our insights into the dynamics of the teacher-student relationship. Recently I heard a talk by Bhanté Henepola Gunaratana that helped me to focus my musings on a method of confronting sexual abuse. Bhanté spoke of *brahmadanda* or shunning. He explained that brahmadanda literally means "noble staff" and that metaphorically it means "noble penalty."[5] In the *Mahāparinibbāna Sutta*, the Buddha says,

> "After my passing, the monk Channa is to receive the Brahma-penalty."
> "But Lord, what is the Brahma-penalty?"
> "Whatever the monk Channa wants or says, he is not to be spoken to, admonished, or instructed by the monks."[6]

Bhanté explained, "Channa had played a role in the earliest part of the Buddha's career as the charioteer who drove him around his father's compound, where he saw old age, sickness, death, and a monk. Later, as a monk himself, Channa took credit for establishing the Buddha's career. He was arrogant about it, making himself obnoxious. After the Buddha gave instruction that he be shunned, Channa appealed to the Buddha to reverse his verdict and the Buddha refused. Channa saw the light, became repentant, and after a probationary period was readmitted to the order."

When I heard this story from Bhanté, it occurred to me that shunning might be an option today in the cases of teachers who abuse their students sexually. This is a think piece, so let me think.

First of all, can we presume to uproot brahmadanda from a traditional text and transform it into a modern tool? We would have to rally around in an unprecedented manner to make it work in our circumstances. The world community of Buddhists is not a sangha or a network of sanghas of the sort that the Buddha established in his lifetime. We are not cohesive, we do not even have a cohesive community of elders, and we don't have a figure with the moral and spiritual authority of the Buddha Shākyamuni. Even the Dalai Lama doesn't say, "OK, Rinpoché, you're busted."

Second, sexual abuse is more complex than the empty boasts of a charioteer. Yamada Kōun Rōshi used to say, "The practice of Zen is the perfection of character." I understand his words to mean that the function of Zen Buddhist practice, and by extension all Buddhist practice, is to personify the Dharma. Students are drawn to Zen practice specifically to attain this perfection for themselves, and they idealize the teacher as one who has attained it—or at least as one who seeks to practice it. When the paragon turns out to be a sham and declares that he is above the Dharma—that you must venerate him separately from the Dharma—then the student can fall with Satan to the bottom of hell.

"I take up the way of not misusing sex." This is the third of the Pañcha Shīla, the Five Precepts that Buddhists vow to follow—all Buddhists: Theravada, Vajrayana, Mahayana, lay and ordained. These five vows are not original with Buddhism but have their roots in Hinduism and in Zoroastrianism. They are perennial guidelines. The *I-hsin chieh-wen*, an ancient commentary associated with the T'ien-t'ai school and used in the Zen Buddhist ceremony of accepting the precepts, has this to say about not misusing sex:

> Self-nature is subtle and mysterious. In the realm of the ungilded Dharma, not creating a veneer of attachment is called the Precept of Not Misusing Sex.[7]

The Buddhist priest, with his neatly pressed robes and his cleanly shaved head, is the embodiment of the ungilded Dharma. As a sexual abuser he not only displays a veneer of attachment but has concealed the Dharma completely with his ignoble exploitation. Buddhist precepts are not commandments, but as guidelines they reveal the Tao. The Buddhist teacher who is also a sexual abuser reveals what the Tao most certainly is not.

Furthermore, the sexual abuser learns to manipulate transference to create an ultimate kind of loyalty. If one of his senior students or board members becomes disaffected—bingo!—she or he is disappeared, and a new, more faithful disciple is slipped into place.

However, this latter problem could be addressed by starting with an expanded kind of brahmadanda. If those who run the teacher's center are blind in their loyalty, then colleagues from other centers and from academia could agree to practice brahmadanda. The colleagues would have less to lose by taking such a position than deeply invested students. At the same time, they are important to the teacher. Without interaction with colleagues, I, for one, would lack some of the inspiration that helps me to grow. Cut off from such support, the teacher in question might begin to face some very unpleasant facts.

This shunning, like that of Channa, should be initiated with open communication. The malefactor should be reached and told, in effect, "Because we are old friends, because I respect your work, and because I can't stand seeing you ruin your life and the lives of others, I have to take a stand. Ordinarily, I would invite you to take part in this conference, but I can't play the role of codependent anymore. Convince me that you have changed your ways, and we can take up where we left off."

If just a few colleagues shun the abusive teacher in this way, students in the sangha might ask why their teacher no longer appears so frequently at conferences of teachers and at academic symposia in the field of his supposed expertise. This kind of open query, along with private expressions of concern through private channels of friendship with senior members, could start a train of cohesive action. The matter could come up in board meetings, a professional interventionist could be invited in, and a process begun to help the teacher to face his depredations squarely.

There are other options. One would be for fellow teachers to direct a letter to senior members or board members. These would be teachers who have interviewed former students of the abusive teacher and have gained a clear picture of his condition. The letter could be worded as a communication that would be made public if a positive response is not forthcoming. This too could lead to the board taking action.

Another option would be for disaffected senior members to gather and exert pressure from an informed, stable position. Still another (in the works now, I understand) would be for senior teachers to come together as a commission to gather evidence and then to convey the objective findings of fact to the concerned board or senior members.

The ensuing process in the concerned sangha could be like intervention in the case of a substance abuser. Through private counseling, the family, friends, colleagues, and employer learn to say to the addict, "We are your family, your friends, your colleagues, your employer—but until you enter a program for treatment and then get regular professional help to maintain a way of sober conduct, we will suspend this relationship."[8] This is, of course, a last resort, when it is clear that the drive to indulge in drink or drugs overrides good sense and decency. It is the end of the road, after a long history of denial, evasiveness, and broken promises.

In the case of the substance abuser, family and friends must stick together in the treatment process. If someone wavers, then the process doesn't work. In the sangha of the sexually abusive teacher, it is likewise important for the senior members and/or the board to be united. This could be especially difficult. The teacher is likely to be defensive to the bitter end. Senior people who understand the need for action might have to labor with companions who want to cleave to their guru. Perhaps some members might have to resign to make consensus possible.

Implicit and explicit in this process should be an acknowledgment on the part of the sangha that this is a teacher who has simply let the fire get away from him. There is nothing wrong with the fire itself. Can it be diverted, admittedly with painful work, to save the teacher, his victims, and the rest of us? One can but try.

This saving, by whatever name, is our first vow. The purpose of the intervention would be the same as Buddha's in dealing with Channa: to encourage the liberation of the teacher, as well as those for whom he has caused trouble. For the teacher especially but also for the sangha, this liberation would be freedom from self-centered constraints to allow full confession and repentance. The truth is confessed, the past is repented, and all beings are liberated. Everything else is extraneous and should be allowed to drop away. Hang the consequences. If being open and honest all around means no more zendō, then no more zendō. "The truth shall make you free."

The confession and repentance process could involve so much personal pain that the teacher might become quite immobilized. This could be a plateau on the upward path, and the teacher should be encouraged not to tarry there. With the help of the sangha and with someone skilled in helping people who suffer from self-betrayal, he can move on. On the other hand, the teacher might not be able to bring himself to consider confession. In such a case, maybe the best solution would be to encourage him simply to retire and to invite in some talented person to help bring the Dharma to life.

If, however, with the support of his students and the interventionist, the teacher can manage to turn himself around, then the sangha can turn around too. Secrets will be brought into the open, conspiracy will become harmony, and years of earnest practice can be salvaged—not only salvaged but made meaningful as never before. There will be no place for codependency. If the teacher must leave, then he will leave with aloha, as we say in our islands. If he can stay, he will simply be, as Senzaki Sensei used to say of himself, an elder brother in the Dharma. The new openness and well-grounded harmony will surely enrich and enhance the practice of all members.

The risk, of course, is that the process could fail. The teacher might simply become more careful, or he might take a position of denial, split off with devoted followers, and set up another center. Intervention is thus a calculated risk and needs to be coordinated with the utmost patience, love, and wisdom.

It may be that nothing will get the full attention of the abuser except a lawsuit. The legal system is increasingly receptive to claims based on sexual harassment in the workplace, and it seems only a matter of time until other types of sexual harassment are given equal recognition. The extremity of a lawsuit could be a remedy in cases where it is clear that the teacher has left the teachings so far behind that no appeal to compassion and ordinary decency will make a dent, and "tough love" is the only option. The traditional Buddhist distrust of the adversarial approach to conflict would thus be set aside.

Whatever the remedy, it must, of course, run in conjunction with therapy for the victims who have been so badly burned. This can be one-on-one treatment but should include group sessions as well. It seems to me that the sangha should support all such treatment financially and stand ready to help in every other possible way. I am always surprised and disappointed when I hear people blame the victim. "She should not have put herself in such a vulnerable position," they say. *Come on!* If one is not vulnerable, no teaching or learning is possible.

Occasionally I meet a student who doubts every word I say. I don't take this personally—people walked away from Chao-chou, even from the Buddha himself. But such invulnerable students are incapable of entering the stream, at least for now. To be vulnerable, to be naive—that is the Tao.

Still, the people in charge of orienting new students on this path should caution them to listen to their feelings. They should assure the inquirers that it is all right to reflect on their reaction to the words or conduct of the teacher and to feel free to say, for example, "Hey, I'm feeling manipulated." If the teacher doesn't listen and respond appropriately, then they should walk away.

It is also important for the senior people to schedule regular sharing meetings, where new and old students alike feel safe to disclose anything about their lives, including any betrayal they might be experiencing. A skilled facilitator can set a tone of safety in just a few sessions, sometimes in just a few minutes.

Of course, this kind of openness won't happen in the centers that concern us. The sexually abusive teacher will do everything he can to prevent anyone from suggesting that the teachers be questioned or from scheduling a sharing meeting. There will be all kinds of doctrinal justification for these taboos. The Devil can cite scripture—he can use precedent and the teachings to justify the pernicious conduct itself: "This is the crazy wisdom of our tradition." "There is no right or wrong." "Everything is empty anyway."

Thus, openness and sharing are modes that should be in place when the new teacher is installed. Once an abusive teacher is in control, then intervention is likely to be the only viable way to proceed.

Some students are likely to find that their practice becomes flat in the extended strain of the intervention process. Many will probably drift away. "There are things time passing can never make come true," as William Stafford has written.[9] One can just do what one can.

There remains the question of how the situation arises in the first place: how is it that the old teacher could be persuaded to convey transmission to an aspirant whose character is fundamentally undeveloped? The answer may lie in the fact that formal leadership positions in Zen have only recently opened to women and, even now, just barely. Male Zen leaders of the past appear to have shut their eyes to the disturbing power of sexuality by failing to address the subject both in Zen literature and in the process of checking the character of their successors.

A look at the literature confirms this. Taking Chinese culture first, among the fifty-five hundred cases listed in my directory of Chinese kōans, I have found just two that acknowledge sexuality, one of them Sung-yüan's question about the vermilion thread, cited earlier.[10]

In Japanese Buddhism, by contrast, sex is acknowledged but trivialized. Our resident monk at Koko An thirty years ago was mildly admonished by his teacher as a "rascal." That teacher was my teacher as well, but I was unable to communicate my sense that his monk might be suffering from an obsession.

Sexual hypocrisy, rather than exploitation, is seen as the primary evil in Japan. Thus the Japanese master who has an affair is criticized, but much stronger disapproval is directed to the teacher who preaches celibacy while keeping a mistress.[11]

The problems are not limited to teachers from Japan. In recent times, we find Korean, Tibetan, and Vietnamese teachers accused of abusing their students. Sanghas with American and European teachers have had to face the issue.

I have heard people try to excuse Far Eastern teachers by suggesting that their monastic training did not permit much interaction with women, and they were not prepared to work with them as students. I find that to be a specious argument. What happened to the basic experience of the "other" as no other than myself? It doesn't matter that the other is superficially different. Without at least a glimpse of the Buddha Shākyamuni's realization that all beings are the Tathāgata, without continuing practice to clarify that glimpse and make it personal, a teacher cannot lead others on the path of wisdom and compassion. He is not grounded—not a teacher in the first place.

The Buddha said, "It is hard to be born a human being. It is hard to meet the true Dharma." So it seems, even in a long-established center, even with earnest practice. It is my hope that we can bring ourselves forth anew from the conflagrations that are still burning people. We know enough. Let's take ourselves in hand, and share.

This is my contribution to the sharing. I look forward to continuing the discussion.

1995

About Money

IN the Buddhist tradition, money is both clean and dirty. It is dāna, the gift, which supports the temple and its monks and nuns. But it is handled very ritualistically, enclosed in white paper, often conveyed on a tray. It must be purified somehow.

Money is the fodder of Māra, the destroyer, who becomes fatter and fatter with each financial deal at the expense of the many beings. However, money can also be a device for Kuan-yin, the incarnation of mercy, whose thousand hands hold a thousand tools for rescuing those same beings.

Both Kuan-yin and Māra function as the Net of Indra. Each point of the net perfectly reflects each other point. Each point is a hologram. Māra says, "All of you are me." Kuan-yin says, "I am all of you." It's the very same thing, except in attitude. Attitude poisons or nurtures the interbeing.

Ta-sui announced that you and I perish along with the universe in the kalpa fire at the end of our eon.[1] Joyous news! Joyous news! Money disappears. Suffering disappears. Even Māra and Kuan-yin disappear in the laughter of Ta-sui. How to find Ta-sui's joy is the question. The path is eightfold, the Buddha said: Right Views, Right Thinking, Right Speech, Right Action, Right Livelihood, Right Effort, Right Recollection, and Right Samādhi.

Māra hates Ta-sui, for he confirms the demon's worst fears and seems to exult in them. How can he joke about the ultimate end! How can he threaten the structure of power and the system of acquisition! Māra hates the Eightfold Path because it undermines the ramparts of his firehouse. The firehouse becomes a hospice and his champion firefighters become nurses. Who will put out the kalpa fire?

Meanwhile Kuan-yin reposes on her comfortable rock by the waterfall, shaded by a willow tree. People say they don't like bowing to Kuan-yin because she is just an icon or an idol. Of course it's nonsense to bow to an idol. Kuan-yin doesn't think of herself as an idol. Her idol is her ideal; her ideal is her Right Views; her Right Views are her blood and guts.

Kuan-yin's practice is elemental too. It is embodied everywhere —as the Earth, for example, exchanging energy with Uranus and Jupiter and Mercury and the others together with the sun as they plunge on course through the plenum. It is embodied as the plenum itself with its incredible dynamics of nebulae and measureless, empty spaces. You will find the dāna of Kuan-yin in tiny systems of mutual support, as well—the termite, for example, nurturing parasites who digest our foundations in exchange for a dark wet place to live.

Primal society also embodies the dāna of Kuan-yin, circulating the gift that nurtures its families and clans. At a single festival, a necklace of precious shells becomes two dozen precious pendants. At a single market holiday, a knife becomes salt and salt becomes a colt. The honor of a new chief is spread by blankets far and wide.

Of course, Māra blows his smoke through these exchanges. Did the primal peoples know Māra from Kuan-yin? They never heard of either, of course, but they knew greed when they saw it and so do we.

Māra isn't an icon either, and he is bowing to himself all day long. He hates the notion of circulating the gift. Instead he circulates the folks. He maneuvers them, lines them up before his machinery, then offers them their products for their money. He circulates the animals and their products, the grasses and their products, the trees and their products. Broken glass set in cement on the tops of high stone walls protects his treasure from those whose diligence produces it. Gates and armed guards and police dogs protect his children, and judges protect his bookkeepers. With his ardent practice, the poor get so poor that he must give a little back to keep the arrangement functioning. Then he is ennobled, and great institutions of benevolence bear his name. Bits of nature are conserved. Peruvian musicians are recorded. Yet wealth can save all beings. Its karma can be inspired by Kuan-yin. The wealthy are stewards named Kuan-yin.

All the while Kuan-yin herself sustains the poor. They are her teacher. She doesn't circulate the folks or their products; she leaves them be. She leaves the birds and the fish and the animals be, the stones and trees and clouds—and does not move them around. The walls with their broken glass and guarded gates hold her in her place, outside. Out there, if she keeps the folks entertained, she might even get a grant. You can have a grant and do your thing, or you can go to jail. It's up to you, Kuan-yin.

It isn't easy for Māra to manipulate people and things. He practices so diligently that he forgoes golf and the theater sometimes. Kuan-yin forgoes golf and the theater too as she sits in royal ease, delighting in the birds as they dip in and out of the spray. But Māra never finds ease of any kind, not even in the middle of the night. His prostate gives him hell, and he sweats with fear.

This is the uneasy, primordial mind, arising from the muck, as reptilian as any dinosaur. It is much older than Kuan-yin. How old is Kuan-yin? Don't say ageless. You are just letting Māra do his dirty work unchallenged. Don't say she is the moment. That is Māra's view as well, pouring out the drinks at his villa on Majorca.

Māra can be your fallback mind and mine as well, always there. Kuan-yin, on the other hand, is eternally fresh and new. She can come into our time and go out of it freely, a trick Māra never learned. We cannot fall back on Kuan-yin; we have to remember her. With a single Māra thought we are in his reins—giddiyap, horsie! With a single Kuan-yin thought, we are laughing at the puppies. Namo Avalokiteshvara Mahāsattva! Namu Kanzeon Makasatsu! Veneration to the Great Being Who Perceives the Sounds of the World!

Māra and Kuan-yin create and cultivate many nets within the Net of Indra. Like the stars, the points in these lesser nets survive by exchanging energy, called money by Māra, called money by Kuan-yin sometimes. There are industries and collectives, golf clubs and base communities. In the lesser nets, Māra dominates, Kuan-yin subverts. Māra co-opts the subversion. Kuan-yin chooses to counter with her money sometimes, if it will keep the waterfall abundant and the birds happy. Sometimes Kuan-yin runs an industry. Sometimes Māra runs a collective. Sometimes there are base communities within golf clubs. Sometimes there are golf clubs within base communities.

It is possible to play endlessly with archetypes and metaphors. Māra as the reptile mind can be called the id. Kuan-yin can be called the superego. When the id is boss, the forests burn in Armageddon's self-fulfilling prophecy. When the superego is boss, the fires of love are extinguished. But Māra and Kuan-yin are not Māra and Kuan-yin; therefore we give them such names. Wipe away the terminology! Wipe away the archetypes! Let Māra and Kuan-yin disappear!

The anguish of nations and families arises from an anxiety to prove oneself—or oneself together with kin and compatriots. The vow to save everybody and everything can bring fun to the dinner table and to international festivals. Proving yourself is the Way of the Buddha, bringing forth your latent pantheon of Mañjushrī, Samantabhadra, and the others as the self. (The archetypes keep popping up anyway!) But remember that the vow to save everybody and everything can be the imperative to bring Mother Hubbards to clothe naked Hawaiians.

Checks, bills, bonds—the tokens of power—transport solutions of sugar and salt to rescue infants from dysentery. They prime the pump of life and order eggplant Parmesan at Auntie Pasto's Restaurant. They build the dam of energy. Moose and beavers and primal people die. Checks, bills, and bonds dance to the music of attitude. Māra has his music, Kuan-yin hers.

We're all in this great mess together. You can't hide out and drink from streams and eat from trees. Or if you do, you are languishing at the top of a hundred-foot pole. Ch'ang-sha will kick you off.[2] The culture we treasure does not exist apart. The municipal symphony, museums, galleries, theaters, bookshops, even our practice centers are intimately integrated into the acquisitive system. We have to work with this fact somehow. It is not clear to me, as it may not be clear to you, how to go about this. As you go along, the qualms can get worse. You can find yourself in a truly dark night, with many misgivings about the Way and doubts about how to deal with the terrible ethical problems that confront everyone—teachers, social workers, managers, homemakers, plumbers, receptionists.

I suggest that the way to deal with a lack of clarity is to accept it. It's all right not to be clear. The practice is to clarify. Moreover, you're working always with your ego. You never get rid of your ego. Your ego is just your self-image. Burnish your ego down to its basic configurations. Then it will shine forth. You can forget yourself as your vows take over your practice, like the birds in the spray of the waterfall.

1993

TAKING PLEASURE

IN THE DHARMA

Herald Birds

The chattering of birds and the humming of insects
are secrets imparted to the heart-mind.
TS'AI KEN TAN[1]

In Western folklore, the ultimate secrets are imparted by angels, winged messengers who have no memory, as Dante says. They come forth *as* their communiqués. Natural phenomena convey secrets to primal people—birds, animals, plants, waters, mountains, meteorological incidents. In Mahayana Buddhist folklore, it was the Morning Star that conveyed realization to the Buddha—a beacon "from the abode where the Eternal are." For Japanese, the *uguisu*, the bush warbler, is a holy messenger. Here is Lafcadio Hearn's experience of that herald bird, which he recorded almost a hundred years ago:

"*Ho—ke-kyō!*"

My *uguisu* . . . is awake at last, and utters his morning prayer. You do not know what an *uguisu* is? An *uguisu* is a holy little bird that professes Buddhism. All *uguisu* have professed Buddhism from time immemorial; all *uguisu* preach alike to men the excellence of the divine Sutra.

"*Ho—ke-kyō!*"

In the Japanese tongue, it is *Ho-ke-kyō*; in Sanskrit, *Saddharma-Pundarika*: "The Lotus of the Good Law" the divine book of the Nichiren sect. Very brief, indeed, is my little feathered Buddhist's confession of faith—only the sacred name reiterated over and over again like a litany, with liquid bursts of twittering between.

"*Ho—ke-kyō!*"

Only this one phrase, but how deliciously he utters it! With what slow amorous ecstasy he dwells upon its golden syllables!

It hath been written: "He who shall keep, read, teach, or write this Sutra shall obtain eight hundred good qualities of the Eye. He shall see the whole Triple Universe down to the great Hell Aviki, and up to the extremity of existence. He shall obtain twelve hundred qualities of the Ear. He shall hear all sounds in the Triple Universe,—sounds of gods, goblins, demons, and beings not human."

"*Ho—ke-kyō!*"

A single word only. But it is also written, "He who shall joyfully accept but a single word from this Sutra, incalculably greater shall be his merit than the one who shall supply all beings in the four hundred thousand Asankhyeyas of worlds with all the necessaries for happiness."

"*Ho—ke-kyō!*"

Always he makes a reverent little pause after uttering it and before shrilling out his ecstatic warble,—his bird-hymn of praise. First the warble; then a pause of about five seconds, then a slow, sweet solemn utterance of the holy name in a tone as of meditative wonder; then another pause, then another wild, rich, passionate warble. Could you see him, you would marvel how so powerful and penetrating a soprano could ripple from so minute a throat, for he is one of the tiniest of all feathered singers, yet his chant can be heard far across the broad river, and children going to school pause daily on the bridge, a whole *cho* away, to listen to his song.[2]

From time immemorial, from time aeons before the Buddha appeared in the world, the *uguisu* has been chanting the praise of the Lotus of the Good Law, the *Saddharma Pundarika Sutra*. What is the Good Law? The Buddha received it from a living point in the early morning sky and decoded it for the rest of us: "All beings are the Ta-thāgata; only their delusions and preoccupations keep them from testifying to that fact."

Western culture too can be porous enough sometimes for winged messengers to impart their secrets. Here is part of a letter from B. N., a student of Zen Buddhism in Sydney, Australia, who was visited by a tiny herald that prepared her for an experience of liberation:

> I'd also like to tell you about an experience I had on Christmas Day. I was sitting on the verandah with S. [her young daughter] when a rainbow lorikeet came down and walked along inspecting the crumbs we offered it and checking out S.'s feet. This was really unusual—they are treetop birds, and I had never seen one at our place before. It was really a Christmas gift to have that contact.
>
> After S. wandered off, I continued sitting here reading Joko's book. I finished reading an essay that ended with something like "no past, no present, no future," and then I wanted to cry. I plunged into alternate crying and laughing—over-whelming and very strong—in which I was aware of the gum trees in the gully in a way I hadn't seen them before. They seemed somehow freed of the past and future I usually see in them. Out of this came words to the effect, "Nothing and no-body needs me, and I'll never stop looking after any of them." This came back to me on and off during the day as some sort of divine joke, and also as a commitment for the rest of my life. I have lost the immediacy of it now, but it is one of the lights that will sometimes gleam through the fog for me. That eve-ning I went up to the lookout in the dark, and remember laughing and crying with the kookaburras. I also saw the bats flying over—the first time for ages that I have seen them, and also a gift, like the lorikeet.

After the lorikeet prepared the way, B. N. came upon a passage in Charlotte Joko Beck's *Everyday Zen* and found herself in a single moment freed of past, present, and future. In the same cosmic turn, the eucalyptus trees too were liberated, and she found herself in the sacred community. She laughed and cried with the kookaburras and was gifted by the bats.

This was a renewal of ancient experience. A winged messenger linked with Dharma teaching brought Pai-chang profound realization in the following well-known story:

> Ma-tsu and Pai-chang were taking a walk. Suddenly a wild duck flew up. Ma-tsu asked, "What was that?"
>
> Pai-chang said, "A wild duck."
>
> Ma-tsu asked, "Where did it go?"
>
> Pai-chang said, "It flew away."
>
> Ma-tsu laid hold of Pai-chang's nose and gave it a sharp twist. Pai-chang cried out in pain.
>
> Ma-tsu said, "Why! When did it ever fly away!"[3]

When indeed! Ma-tsu's bulletin of interbeing and Pai-chang's realization could only have occurred with the wild duck appearing from nowhere, at that particular juncture of holy opportunity. The story continues:

> After the two worthies returned to their temple, a monk found Pai-chang weeping in his room. "Why are you weeping?" he asked.
>
> "Go and ask the teacher," replied Pai-chang.
>
> The monk asked Ma-tsu about it, and Ma-tsu said, "Go back and ask Pai-chang."
>
> The monk returned to Pai-chang and found him laughing. "A while ago you were weeping," said the monk, "and now you are laughing. Why is that?"
>
> Pai-chang said, "A while ago I was weeping and now I am laughing."[4]

Pai-chang's final words form one point. Another is, I suspect, that when Pai-chang seemed to be weeping, he was really laughing, and when he seemed to be laughing, he was really weeping, like B. N. who laughed and cried in authentic response with all beings.

Here is another account of a bird and its sacred gift, set forth by C. D., a student living in Honolulu, who titles his piece "Meeting Pueo."

This was my eighth or ninth trip up Mauna Loa volcano. I usually go once a year, in the late summer. The sparse cover of *ohelo* bushes, fountain grass, and 'ōhia trees gives way to barren lava as the trail gains in altitude. Just before the eight-thousand-foot marker there is an 'ōhia tree, about twelve or fifteen feet tall, growing up out of an old pahoehoe flow. There is no other vegetation within a mile or so that is even close to the size of this tree; its spare but graceful form is a natural landmark on the trail. There are no trees upslope. We nicknamed it "Last Tree."

On our return from the summit my hiking companions and I decided to take a break at Last Tree. Someone noticed a pueo [Hawaiian owl] coming our way. It fluttered up the slope like a large butterfly, pumping up toward us from down below in the clear morning air, then stalling, hovering, then continuing, then hovering. At one point it spied us, stopped in midair as if to size us up, then turned and flew away. The three of us were rapt under the 'ōhia tree. Eventually, my companions loaded up and headed down the trail.

When they were gone I had a chat with the tree, saying thanks, telling it that I would be back again next year, and then gave it a good firm embrace. As I shouldered my pack I thought about making a sketch of Last Tree. I had carried a sketchbook with me up the mountain but here it was the fifth—and last—day of the hike and I had not yet used it. We were making good time down the mountain and it was hard to break the momentum but I stopped and unshouldered my pack to the side of the trail below the tree and fished out my sketchbook. 'Ōhia blossoms spread out like outstretched fingers into the crisp blue sky, the volcano's mass pushing up the broad curve of the far horizon.

Suddenly I heard a quick WHUFF-WHUFF-WHUFF and saw a shadow on the rocks just in front of where I was kneeling. A small surge of adrenaline rushed through my veins, nerves wide open from five days of mountain solitude. I tilted my head back and saw the pueo hovering a dozen feet above me—two black eyes set into a white downy visage. The owl was close: I could feel the wingbeat, see the feathers. Then pueo angled into a breeze that carried it swiftly downslope and it was gone.

I felt a strange sense of power and elation, that something fortunate had transpired, and that the best way to accept the gift was to keep acting as if nothing much had happened. I finished the sketch and hiked a couple hours to cold beers at the trailhead brought by the friend who came to pick us up.

It is since meeting pueo that I have noticed something different in my circumstance and bearing, a sense of confidence and trust. Before this meeting with pueo it was somehow easier to indulge in doubt and despair; now I don't feel as prone to those states. In place of doubt or the feeling of a general absence of *toku* [Japanese: "power, virtue"], I feel trust and calm, that there is a spirit companion that will help with whatever it is that needs to get done.

This sense is not entirely new; I've had these feelings before. They seemed to come and go in some sort of cycle, as do human relationships. Now, however, I have a better handle on the coming and going. If I'm worried or doubting, an image of pueo might flash through my mind, or I will "hear" a bird sound and it will seem as though pueo is close, and I get a feeling of fearlessness. It is as though I have a friend who is looking out for me in some way that I only dimly perceive. The rather dreamlike image that concocts itself in zazen or when I'm just thinking about pueo is that of pueo (the totem) gliding over the broad expanses of lava on the flanks of Mauna Loa (emptiness, the empty universe). Pueo flies away, over the lava, deeper and deeper into emptiness; that is the image I've had since the encounter that is most compelling.

C. D.'s mysterious and heartening communication with the pueo is validated in Hawaiian folklore, where it is a bird of guidance. It is also validated generally in Mahayana teaching. The Sambhogakāya for the traditional Buddhist is the Body of Bliss, the bliss of harmony and the most intimate communication. What is that communication? It cannot be expressed in words, but it verifies what it is to be a being among beings.

Finally, here is the account of T. L., born and raised on the Big Island of Hawaiʻi, of his communication from herald birds:

The Volcano House perches on the rim of the vast Kilauea Caldera on the island of Hawaiʻi. Within Kilauea, the circular Halemaʻumaʻu Crater sinks away from the floor like a mysterious Hawaiian cenote. Here, Pele, the Hawaiian goddess of fire and the volcano, is said to make her home, receptive to offerings of gin and other more traditional sacrifices.

More than just a rustic lodge, the place of the Volcano House as the chief repository of Pele lore was ensured by its longtime owner, the colorful Greek, George Lycurgus. "Uncle" George, who was reputed to be Pele's lover, lived to be over one hundred years old. During his long tenure he entertained famous guests from all over the world, and could often be seen playing his favorite game of cribbage in the hotel lobby.

The volcano region is encompassed within the district of Kaʻu, a wild windswept place whose people are stoic, proud, and uncommonly rooted in their land and traditions. I was born in the town of Pahala in the middle of Kaʻu. My family moved to Hilo, a small city fifty miles away, when I was very young, but my affinity for Kaʻu remains very close—something I cannot explain.

In the course of becoming an adult I moved to Los Angeles and lived there for nearly ten years. I then returned to Hawaiʻi and got a job as night auditor of the Volcano House. "Uncle" George had already died. When I finished with my duties as auditor, I would sit before the fire made famous in Ripley's "Believe It or Not" as never having gone out in a hundred years. As the only person on duty for much of the night, it was my job to see that the fire kept burning. Several times, after having dozed off, I would awaken to find the fire reduced to glowing embers and I would urgently stoke it back to life, adding more logs until it was a roaring blaze again. It was during these early morning hours that I had my first encounter with a mute winged messenger whose message, if there was one, remains a mystery to me.

The koaʻe kea (white-tailed tropic bird) is a large white oceanic bird with two very long tail feathers that because of their length, flutter like paper streamers when the bird is in flight. The wingspan of the mature bird is almost three feet.

I saw koaʻe kea on the island of Kauaʻi where there are many tall cliffs with rocky ledges for nesting, and I knew also that they nested within the Kilauea Caldera. Though far inland, Halemaʻumaʻu Crater with its vertical sides and rocky ledges also provides good nesting sites. The koaʻe kea is not a *kino lau* (alternate form) of Pele, but in my mind they are inextricably mixed.

One dark morning, instead of sitting before the fire as I usually did, I sat next to the window facing the caldera. Suddenly a koaʻe kea crashed into the glass. It fluttered against the glass for a few moments, then abruptly flew away. Though I reasoned that it was probably attracted to the light, I felt an immediate connection, as though it were trying to communicate something to me.

I worked at the Volcano House for five years, commuting nightly from Hilo. I then decided to go back to school to earn a teaching certificate. Often I went down to the beach for a swim. One day, I was sunning on the rocks and saw a large white bird flying toward me. As it got closer, I recognized it as a koaʻe kea with its unmistakable long tail feathers fluttering behind. I was startled to see it so far from Kilauea. It flew directly overhead, circled three times and flew off. In thinking back on the event, I recall that I wished that it would return and come closer. It did, close enough for eye contact, and I felt a definite unspoken communication between us.

A couple of years later, when I had already started teaching, I returned home from work one day and drove into the garage. (My home is on Halai Hill, a place associated in mythology with Hina Keahi, another goddess in the Hawaiian pantheon.) As I walked from the garage, I turned the corner of the house to get to the front door when I saw four koaʻe kea flying around the front lawn, which is smaller than a tennis court. The sight of these large, snow-white birds flying before me was so startling that I thought I was dreaming. They flew

higher and higher, frolicking in the air, for several minutes before they flew off.

At first I interpreted these unusual encounters as portents,
but no unexpected events occurred within a reasonable time
for a connection to be made. Perhaps the desire to read meaning into these encounters is an unwarranted intrusion by the
rational mind. However, I do not believe that the awe, the
sense of mystery, and communion I felt during those transfigured moments have been diminished as a result.

I have told somewhere about a little fox that played with me in the
hills above La Crescenta, California. Years later, sitting quietly in the
forest at San Juan Ridge, day after day in the same place, I found that
animals and birds would show themselves to me and even bring their
children. Such occasions, to use T. L.'s terms, were transfigured moments when the ordinary sequence of thoughts falls away. Readied
by practice and certainly nurtured by a naive spirit, then the lorikeet,
the wild duck, the pueo, and the koaʻe kea will have a chance to convey their messages, cutting the loop for all time and transforming
the thinker.

1993

Wallace Stevens and Zen

I THINK it would be fair to say that certain Asian vapors were part of Stevens's Hartford, but they were faint. He had a Buddhist image in his room, sent by a friend from Ceylon, which he liked because it was "so simple and explicit" (L., 328).[1] He admitted to influence by "Chinese and Japanese lyrics" in one letter and denied the importance of such influence in another (L., 813 and note). Buddhism itself is not mentioned once in his letters, unless we count a passing reference to "Buddha and Christ" (L., 632).

Nonetheless, there is a profound relationship between Stevens's work and the teachings of Zen Buddhism. The ground of this relationship is "a mind of winter," where there is no intellectual overlay to obscure things as they are:

> One must have a mind of winter
> To regard the frost and the boughs
> Of the pine-trees crusted with snow;
>
> And have been cold a long time
> To behold the junipers shagged with ice,
> And spruces rough in the distant glitter
>
> Of the January sun; and not to think
> Of any misery in the sound of the wind,
> In the sound of a few leaves,

Which is the sound of the land
Full of the same wind
that is blowing in the same bare place

For the listener, who listens in the snow,
and, nothing himself, beholds
Nothing that is not there and the nothing that is. (9)

The title of this poem, "The Snow Man," refers not to a construction of snow with two pieces of coal for eyes but rather to a person who has become snow. A snowman is a child's construction; a Snow Man is a unique human being with "a mind of winter," or, as Yasutani Haku'un Rōshi used to say, "a mind of white paper." Look again at Tung-shan for this mind:

A monk asked Tung-shan, "When cold and heat visit us, how can we avoid them?"
Tung-shan said, "Why not go where there is neither cold nor heat?"
The monk asked, "Where is there neither cold nor heat?"
Tung-shan said, "When there is cold, let the cold kill you. When it is hot, let the heat kill you."[2]

"Killed with cold" is to "have been cold a long time." That is the place where there is neither cold nor heat as a concept. When it is cold, one shivers. When it is hot, one sweats. There is just cold or just heat, with no mental or emotional associations "in the sound of the wind, / In the sound of a few leaves."

The ultimate experience of perception of "pine-trees crusted with snow" or of "the sound of the wind" is the explicit sense that there is only that phenomenon in the whole universe; as Stevens expresses it, "the sound of the wind . . . is the sound of the land." This is the nature of seeing or hearing for the Snow Man, perception by the self that has been killed with cold. It is the mind of white paper that is confirmed by that sight, that sound. Dōgen wrote, "That the ten thousand things advance and confirm the self is enlightenment."[3] In other words, it is that form, that sound, which make up my substance. "I am what is around me" (86). Look again at Yün-men's words about the self:

> Yün-men said to his assembly, "Each of you has your own light. If you try to see it, you cannot. The darkness is dark, dark. Now, what is your light?"
>
> Answering for his listeners, he said, "The pantry, the gate."[4]

In maintaining a mind of winter, Yün-men finds his light. There is nothing to be called the self except its experience of the pantry, the gate, and "the junipers shagged with ice." " 'The soul,' he said, 'is composed / Of the external world' " (51).

But when the mind is sicklied over with concepts of the wind as a howling human voice, then also clouds are faces, "oak leaves are hands" (272), and the self perversely advances and confirms the ten thousand things. This is projection, the opposite of true perception, and is, as Dōgen says, delusion[5] — the fantasy of Lady Lowzen, "for whom what is was other things" (272). As Ching Ch'ing says, "Ordinary people are upside down, falling into delusion about themselves, and pursuing outside objects."[6] Presuming that our emotional concerns are the center, we project ourselves onto the wind and the leaves, smearing them with our feelings. We have not yet reached the place where there is neither cold nor heat. We fall into delusion about ourselves and seek to enlarge that delusion by the pathetic fallacy. Stevens had great fun mocking such self-centered fantasy:

> In the weed of summer comes this green sprout why.
> The sun aches and ails and then returns halloo
> Upon the horizon amid adult enfantillages. (462)

"Enfantillage" means child's play or childishness. "Adult enfantillages" I would understand to refer to the ascription of human qualities to nonhuman things, beginning with "why," the conceptual weed that takes us furthest from realization of things as they are, and continuing with the projection of aches and other silly business on the sun. This is the imagination that is not grounded in a mind of winter and is thus infantile.

Vital imagination has its roots in the bare place outside, which is "the same bare place / For the listener"—a generative, not a nihilistic, place. Yamada Kōun Rōshi says, "The common denominator of all things is empty infinity, infinite emptiness. But this infinite emptiness is full of possibilities."

Empty infinity and great potential, the nature of all things as realized by the mature Zen Buddhist—this is also the vision of the Snow Man, with his mind of winter and his capacity to perceive phenomena vividly. Indeed, the final line of "The Snow Man," "nothing that is not there and the nothing that is," precisely evokes the heart of Zen teaching:

> Form is no other than emptiness,
> Emptiness no other than form.[7]

This emptiness of all phenomena, including the self, is being uncovered in modern physics. What appears paradoxical emerges as the complementarity of the suchness and emptiness of all things. This the mind of winter perceives.

Dōgen expressed this complementarity in experiential terms:

> Body and mind fall away!
> The fallen-away body and mind![8]

When body and mind fall away, the self is zero. The listener is "nothing himself" and thus experiences the "nothing that is not there," which is all things just as they are, with no associations—just "the junipers shagged with ice." With this perception, the great potential is fulfilled, and all things are the self: "The fallen-away body and mind!" That is the self as white paper filled with the sound of the wind and the sound of a few leaves.

Bodhidharma, who brought Dhyāna Buddhism from India to China and is revered as the founding teacher of Zen, conveyed this same teaching:

> Emperor Wu of Liang asked Bodhidharma, "What is the first principle of the holy teaching?"
> Bodhidharma said, "Vast emptiness, nothing holy."[9]

"Vast emptiness" is not only the common denominator of all things, it is itself all things, all space, all time together. As Wu-men wrote:

Before a step is taken, the goal is reached;
Before the tongue is moved, the speech is finished.[10]

Stevens wrote, in "Notes Toward a Supreme Fiction":

There was a muddy centre before we breathed,
There was a myth before the myth began,
Venerable and articulate and complete. (383)

This is as far as one can trace Stevens's credo as set forth in "The Snow Man," but "Tea at the Palaz of Hoon," the companion of "The Snow Man" at their first publication, is, I feel, its completion (65). Hoon's descent "in purple," with its connotations of royalty, represents the kinglike nature of one who emerges from emptiness, like the Buddha rising from his profound experience under the Bodhi tree.

I was the world in which I walked, and what I saw
Or heard or felt came not but from myself;
And there I found myself more truly and more strange. (65)

When I was a young lay student in a Japanese Zen monastery, I was surprised at the way the monks would seem to equate confidence with religious realization. Their dignity was regal when genuine, merely arrogant when false, but in both instances, it was quite contrary to the humble attitude I had previously associated with religion.

Stevens knew better. He would have appreciated D. T. Suzuki's translation of a line by Wu-men: "In royal solitude you walk the universe."[11] Professor Suzuki took liberties in using "royal" in this instance, for it does not appear in Wu-men's original Chinese.[12] I feel sure that he was projecting his own experience of empty potency here and that he shared the vision of "mountain-minded Hoon" (121). Fully personalizing "the junipers shagged with ice" is to realize those junipers are none other than myself. "I was the world in which I walked" (65). Confirmed by all things, Hoon walked the universe in royal aloneness, "And there I found myself more truly and more strange" (65). One is reminded of words attributed to the baby Buddha immediately after his birth, which Zen teachers are fond of quoting:

> Above the heavens, below the heavens,
> Only I, alone and revered.[13]

Thus in different epochs and in different cultures, Wallace Stevens and Bodhidharma and his successors present the potent emptiness. I do not know how this could be, but there it is, perhaps no more remarkable than that they had the same number of sense organs. As Nakagawa Sōen Rōshi once said, "we are all members of the same nose-hole society."

But I think we have here something far more significant than human beings expressing common humanity. We are touching the connection between a certain kind of poet and a certain kind of religion. Zen teachers from the very beginning peppered their discourses with quotations from such poets as Tu Fu and Bashō, poets who had little or no formal connection with Zen. Of course, Zen was a part of the cultural atmosphere of T'ang China or Tokugawa Japan, but Tu Fu and Bashō were no more "Zen poets" than Stevens was. It is here that "the green sprout why" would take over our cultivation if we let it. I am content to acknowledge Stevens as one of the very few great poets who will be a source of endless inspiration for future generations of Western Zen teachers.

1980

Play

LIN-CHI asked Huang-po: "What is the clearly manifested essence of the Buddha Dharma?" Huang-po hit him.
This happened three times.

Lin-chi then went to Ta-yü. Ta-yü asked, "Where did you
come from?"

Lin-chi said, "From Huang-po."

Ta-yü said, "What does Huang-po have to say?"

Lin-chi said, "I asked him three times, 'What is the clearly
manifested essence of the Buddha Dharma?' and he hit me
three times. I don't know whether I was at fault or not."

Ta-yü said, "Huang-po is such an old grandmother. He
completely exhausted himself for your sake. And you come
here asking whether or not you were at fault!"[1]

With this Lin-chi had great realization, and exclaimed,
"Ah, there is not so much to Huang-po's Buddha Dharma!"

Ta-yü grabbed hold of Lin-chi and said, "You bed-wetting
little devil! You just finished asking whether you were at
fault or not, and now you say, 'There isn't so much to Huang-
po's Buddha Dharma.' What did you just realize? Speak,
speak!"

Lin-chi jabbed Ta-yü in the side three times. Shoving him
away, Ta-yü said, "Huang-po is your teacher. It's not my
business."[2]

A lot can be said about this case, but I just want to take up a single point. How much is "not so much"? How is it that "not so much" gave rise to such a vigorous tradition that thrives to this very day?

Of course, Lin-chi was not the only teacher in our lineage who talked about the poverty of the Buddha Dharma. The literal meaning of Chao-chou's "Mu" points to the same fact,[3] yet according to the *Book of Serenity*, the monk went on to ask Chao-chou, "All beings have Buddha Nature, how is it that the dog has none?" Chao-chou said, "Because of its inherent karma."[4]

Karma and Buddha Nature, the substantial teaching of all the Buddhas and its empty content—these sets of relative and absolute, the universe and the void, are one in our play as Zen students, thanks to our marvelous heritage.

Huang-po, Ta-yü, Chao-chou, and all the other great ones fooled with themes of essence and phenomena to enlighten us. One of my early Japanese teachers and I used to argue about "play." His understanding of English may have been a factor in our disagreements. For him, play was limited to children, baseball, and theater. I understood play as the nature of interaction—not only human interaction but all of it. Puppies are more frisky than dogs, but even an old dog knows it's a game.

Interaction is play because it doesn't amount to much, or even to little. On your cushions in the meditation hall, nothing impedes your interaction with thoughts. You view one thought-frame after another. When your thoughts wander and you notice what has happened, then easily and smoothly you return to focusing on Mu. When the bell rings for the end of the period, you bring your hands together, rock back and forth, swing around on your cushion, and stand up.

In the workaday world, again, interaction is play. Nothing impedes your response to your child's demands. When the telephone rings, you type "save" on your computer, pick up the receiver, and say, "Hello." When the bus reaches your station, you get off promptly.

> Farmers sing in the fields
> Merchants dance at the market.[5]

Layman P'ang wrote:

> How wonderful, how miraculous!
> I draw water, I carry kindling![6]

When Joanna Macy and I spoke at a Buddhist Peace Fellowship meeting in Sydney recently, we were challenged from the back of the hall by a group of evangelical Buddhists. Are you surprised that there could be evangelical Buddhists? Evangelism is a character trait and is not limited to any particular religion. These people were born-again Buddhists, firmly convinced that "Dharma" and "karma" are entities with certain fixed qualities and tendencies. Joanna and I told them, each in our own way, that no concept is solid or absolute and that even "Buddha" self-destructs. Their Dharma was not ours. They became angry because they didn't know our interaction was play, an inning in the joyous game of time and space, giving and taking with empty universal nature.

"All the world's a stage." We play roles: Zen teacher, Zen student, parent, spouse, friend, worker, pedestrian, and so on. We play "as if," to use the Hindu term, as if we were Zen teacher, student, parent, and so on. The child plays house, as if she were a mother. The mother plays house in exactly the same way.

> He himself took the jar
> and bought wine in the village;
> now he dons a robe
> and makes himself the host.[7]

And when the play doesn't make you laugh, that doesn't mean it isn't play anymore. Tragedy is play too — tragic to the very bottom, perhaps, but still play.

The Knight of the Burning Pestle, by Francis Beaumont, taught me that the audience creates the play, and the play is not confined to the stage. A druggist and his wife are patrons of the theater, and she doesn't like the way the play begins. She stands up in the audience and starts directing things. Her paramour, the druggist's apprentice, is introduced as a new character, the Knight of the Burning Pestle, with a pestle in flames inscribed as a crest on his shield. We then have a new play, and the separation between audience and actors is broken

down. The inner fantasy of the druggist's wife is acted out onstage, and thus inner and outer too lose their barrier.[8] This is only possible because matter is insubstantial, and there is not a speck of anything to interfere with our complete interpenetration.

In the world of play, a druggist's apprentice becomes a knight, a child becomes a father, a dog becomes a baby, and the insurance agent, throwing off his worries about declining sales, transforms himself into a prince and seduces his tired wife and the mother of his brood, who in turn becomes a ravishing, masked beauty at a mummers' ball.

> In a well that has not been dug,
> water from a spring that does not flow is rippling;
> someone with no shadow or form
> is drawing the water.[9]

This is Zen play. Where is the person with no shadow or form? On the stage of the interview room, you dance your response.

That person with no shadow or form inhabits a dream world that is no other than this world. Traditional people confirmed their dreams in this world with ceremonies and then reentered the dream world again by reenacting their ceremonies. We do the same with our ceremonies. We dedicate the merit of reciting our sūtras to our ancestors in the Dharma and to our parents and grandparents who have died. Are they listening? Of course they are. Nakagawa Sōen Rōshi once exclaimed, "Of course there are Bodhisattvas and angels living up in the sky!"

This is all possible because there is not much to Huang-po's Buddha Dharma, or to anyone else's for that matter. And as to the Buddha Nature of the dog, or of you or me: "Mu!"

1986

Uphill Downhill

Nonverbal Presentations in Zen Buddhist Practice

Dōgen Kigen Zenji often said that in Zen Buddhism, practice and realization are the same.[1] This may not be immediately clear. Looking at such representations of the practice as the Ten Ox-herding Pictures, the sequence to realization is plain enough. What, then, is the unity of practice and its goal?

The point is that you practice the goal, just as you might practice the goal in law or medicine or music. It is your realization to practice searching for the ox, then get a glimpse of it, lasso it, tame it, ride it, turn it loose, and forget about it. You find there is nothing at all, then enjoy again the beauty of the world, and ultimately enter the market-place as the Bodhisattva Pu-tai (Hotei) to the delight of everybody.[2]

This progression sets forth the temporal path of practice, which has no shortcut. In the final ox-herding picture, Pu-tai carries a great sack filled with goodies for children. He mingles with publicans and prostitutes and enlightens them all. Yet Pu-tai's realization is full and complete throughout the sequence, just as yours and mine are, even on entering the dōjō for the first time. It is Pu-tai who begins the search in the first picture, glimpses traces in the second, and catches sight of the ox in the third. Zen practice is not a matter of becoming somebody else.

You are not limited to yourself, however, nor is Pu-tai. In Chinese Buddhism, Pu-tai is identified as Maitreya, the future Buddha. Mai-

treya is doing zazen there in the Tusita Heavens, waiting to be born in this world. Some early Chinese Buddhists worshiped Maitreya, and even today you can find their descendants. We can, however, also understand Maitreya to be you and me being born at every moment, all the while in Tusita. As Pu-tai, Maitreya manifests his realization in this world of time and suffering. He brings gifts to all children, all people, all animals, plants, and things. And not just in a single incident or on a single time line but in the spread of the Buddha Way—the turning of the Dharma wheel through the whole universe in every dimension. In Buddhist iconography, Maitreya can hardly be distinguished from Pu-tai, the fat, jolly "Laughing Buddha."

Still, Pu-tai in the first of the ox-herding pictures is having a hard time:

> Exhausted and in despair, he knows not where to go.
> He only hears the evening cicadas, singing in the maple-
>   woods.[3]

It's an uphill trek, but after long practice through many phases, the herder—"she" as well as "he"—reaches the downhill path into the city. Thus, "uphill" can refer to mustering the self and practicing earnestly in the context of private and social obstacles. "Downhill," on the other hand, can refer to the world mustering *with* the self to turn the wheel of the Dharma together with all beings. "Downhill" may seem more comfortable than "uphill," but both are practice.

Bodhidharma, facing the wall of his cave in zazen until his eyelids fell off and he lost the use of his legs, is the herder, searching with single-minded discipline for deepest realization.[4] However, Bodhidharma is, of course, also the kindly patron of successful enterprise. Bodhidharma and Pu-tai in their various incarnations mingle their attributes, yet they are distinctive fellows, offering differing kinds of inspiration in their particular ways.

There are other archetypes for "uphill" and "downhill" in Zen Buddhist teaching. The two ways of inscribing the *enso*, the circle, would be an example. Most masters inscribe it from the bottom, around clockwise, in the apparent direction of the sun—seemingly the natural way—together with everyone and everything. Some, notably Suzuki Shunryū Rōshi, begin at the top and inscribe the circle around to the left, against the grain, "uphill," in the face of the many obstacles of greed, hatred, and ignorance.

The circle is also found in Zen literature. Our ancestors in the Dharma drew circles on the ground and in the air and circumambulated their teachers.

> When the brother monks Nan-ch'üan, Kuei-tsung, and Ma-ku were on their way to pay their respects to the National Teacher Chung, Nan-ch'üan drew a circle on the ground as a challenge to the others. Kuei-tsung seated himself in the center of the circle, and Ma-ku bowed to him.[5]

Nan-ch'üan laid out the temple with his circle, and Kuei-tsung then erected it to the last detail. Ma-ku then bowed with village worshipers and with the rest of us as well. With no one to witness and everyone to remember, these old brothers danced primordial configurations as their own.[6]

The circle is also the path of circumambulation, a practice found in many ancient cultures. In both Asia and Europe, the usual path is clockwise, as in kinhin, our "walking sūtra" around the dōjō between periods of zazen. The counterclockwise way is unnatural, and in Christian countries is akin to the Black Mass, in which the Latin words are recited backward. The devil is evoked, and you are on dangerous ground. The Scottish term for contrary circumambulation is *withershins* or *widdershins*—literally, "against the way."

I remember that in the early days of Koko An, we used to keep the Royal Palm to our left when we walked to the cottage in the front garden. This permitted us to pass each other without eye contact while going to and from dokusan. William Merwin pointed out to us that we were walking widdershins, and considering the life-and-death nature of our practice, we might be monkeying with forces beyond our control. It was just as easy to keep the tree on the right, so we changed the ritual accordingly. There were, it must be confessed, a few mummers of "Rank superstition!" But where does primordial orientation come to an end and superstition begin? I am not sure, and I'm not taking any chances.

A symbol as old as the circle is the swastika. It is like Suzuki

Rōshi's circle, a kind of wheel that turns against the conventional way of things. Though its opposite, the *sauvastika*, which turns to the right, downhill, is the form most usually found in Buddhist art, you can find both in Sōtō lineage documents and probably elsewhere in Buddhist metaphysical representations.[7] Both are Chinese ideographs meaning ten thousand—by extension, "complete" or "full," implying all the virtues. In India the sauvastika was sometimes written in a form said to resemble a curl, as Vishnu's (and then the Buddha's) breast curl. It is also one of the auspicious signs in the footprint of the Buddha.

In Europe, both the sauvastika and the swastika are called the gammadion ("guh-*may*-dee-un"), a word of Greek origin that refers to their forms as combinations of the capital letter gamma (Γ). In English they are also called the fylfot, which may simply refer to the use of the symbols to fill the foot of a painted glass window. The sauvastika in particular is found in Celtic and other ancient stone remains across the world—Europe, Asia, and the Americas, some dating back perhaps ten thousand years.

I read somewhere that Adolph Hitler was warned that the reverse of the sauvastika—the swastika—was very bad luck, but he decided for his own reasons to adopt it as a Nazi symbol anyway. Today, spiky variations are displayed by neo-Nazis in Idaho and in South Africa. However, such association can only be temporary and can be passed over in our research into the fundamental meanings.

In Zen Buddhism the fundamental meaning of the swastika would be found in the decision of Bodhidharma to abandon the court of South China. He made his way to the northwest, where he took up rigorous practice in a cave behind an abandoned temple. He paid no attention to anybody until an earnest monk cut off his own hand by way of expressing his sincere desire to become a disciple. Both took a pretty steep path.

The sauvastika has its meaning with Pu-tai. He is like a Chinese Santa Claus, but not only is his bag full of candy and toys, his great belly contains the whole universe, as Yamamoto Gempō Rōshi used to say.

If, however, you were to inscribe the sauvastika on a piece of rice paper and then turn the paper over, the swastika would appear—the mirror image of the sauvastika showing through. The two presentations are the same and express the same practice, uphill and down. Turning against the ordinary current of greed, hatred, and ignorance is the way of the Bodhisattva, taking joy in songs of birds and human beings—and responding with compassion to the sounds of their anguish. Our task as laypeople is to live by this piece of rice paper with its single symbol of the Buddha Way, and to leave the world without leaving it. This is the attainment of the Way, the highest reach of practice.

Yang-shan and a brilliant monk danced this attainment—the ensō both ways, and then the sauvastika, to the delight of the assembly and generations of Zen students to our present time:

> A monk asked Yang-shan, "Do you know ideographs?"
>
> Yang-shan said, "Enough."
>
> The monk circled Yang-shan to the right and asked, "What ideograph is this?"
>
> Yang-shan drew the graph for "10" in the air [in Chinese this is like the mathematical *plus*-mark, and implies "complete"].
>
> The monk circled his teacher to the left and asked, "What ideograph is this?"
>
> Yang-shan modified the "10" to make a sauvastika.
>
> The monk then drew a circle in the air, and lifted his arms like a titan, as though holding the sun and moon, and asked, "What ideograph is this?"
>
> Yang-shan drew a circle around the sauvastika.
>
> The monk then posed as a guardian deity.
>
> Yang-shan said, "Right, right. Guard it with care."[8]

These marvelous teachers are like Nan-ch'üan dancing with his brother monks while on pilgrimage, presenting primordial orientations as themselves within the boundless, formless, universal circle. It is said that in the early days of Ch'an, or Zen Buddhism, there was a system of circular presentations that might be compared to Tung-shan's scheme of Five Modes of the Particular and the Universal.[9] It was passed by the National Teacher Chung through his disciple Tan-yüan to Yang-shan. Thomas and J. C. Cleary tell how Yang-shan read the text and then burned it. It was the only copy, and Tan-yüan was upset when he heard about its destruction. Yang-shan then re-wrote the work from memory and sent it back to Tan-yüan. How-ever, in the course of the generations, the book was lost again.[10]

Never mind. The dances of the old teachers are its best explica-tion. Yang-shan and his monk, Nan-ch'üan and his brothers, and our many other ancestors play out their circles and gammadions to the ultimate, and perhaps that's enough symbolism for now. It's time to step forth from the dōjō with Yang-shan's timeless injunction ring-ing in our ears, "Guard it with care!"

1983

Notes

NYOGEN SENZAKI

1. Nyogen Senzaki, "An Autobiographical Sketch," in *The Iron Flute*, trans. and ed. by Nyogen Senzaki and Ruth Strout McCandless (Rutland, Vt.: Charles E. Tuttle, 1961), 161.

2. Senzaki, "On Buddha's Images," in *On Zen Meditation* (Kyoto: Bukkasha, 1936), 99.

3. Senzaki, "On Buddha's Images," 99.

4. In these essays I use the Japanese order in rendering surnames first, except for D. T. Suzuki, Nyogen Senzaki, and others whose names are known in the Western style of surname last.

5. "Sōen Shaku on Nyogen Senzaki" (excerpt), in Senzaki and McCandless, *The Iron Flute*, 159–60.

6. Senzaki, "Sangha," in *On Zen Meditation*, 69. See Friedrich Froebel, *Pedagogics of the Kindergarten*, trans. by Josephine Jarvis (New York: D. Appleton, 1932).

7. I have also heard that Senzaki simply showed up uninvited.

8. Senzaki, "Realization," in *Namu Dai Bosa: A Transmission of Zen to America*, ed. by Louis Nordstrom (New York: Theatre Arts Books, 1976), 31.

9. Nyogen Senzaki and Ruth Strout McCandless, *Buddhism and Zen* (San Francisco: North Point Press, 1987), 26.

REMEMBERING SŌEN RŌSHI

1. Robert Aitken, *Taking the Path of Zen* (San Francisco: North Point Press, 1982), 115.

2. Wallace Stevens, *The Collected Poems of Wallace Stevens* (New York: Random House, 1982), 52.

3. See "The Virtue of Abuse" in this volume.

REMEMBERING BLYTH SENSEI

1. R. H. Blyth, *Zen in English Literature and Oriental Classics* (Tokyo: The Hokuseido Press, 1942, 1958, 1993).

2. R. H. Blyth, *Haiku*, vol. 1 (Tokyo: Kamakura Bunko, 1949); vols. 2–4 (Tokyo: The Hokuseido Press, 1950–52, 1966). *Senryu: Japanese Satirical Verses* (Tokyo: The Hokuseido Press, 1949). *Buddhist Sermons on Christian Texts* (Tokyo: Kokudosha, 1952; South San Francisco: Heian International Publishing Co., 1976).

3. D. T. Suzuki, *Essays in Zen Buddhism: First Series* (London: Luzac, 1928). Wei-lang is the Cantonese pronunciation of Hui-neng. See Wong Mou-lam, trans., *The Sutra of Hui-neng*, published with A. F. Price, trans., *The Diamond Sutra* (Boulder, Colo.: Shambhala, 1974).

4. R. H. Blyth, *Zen and Zen Classics*, vol. 4, *Mumonkan* (Tokyo: The Hokuseido Press, 1966).

OPENNESS AND ENGAGEMENT

1. Filmer S. C. Northrop, *The Meeting of East and West: An Inquiry Concerning World Understanding* (New York: The Macmillan Company, 1946).

2. Thomas Cleary and J. C. Cleary, trans., *The Blue Cliff Record* (Boston: Shambhala, 1992), 566–67.

3. Daisetz Teitaro Suzuki, *The Training of the Zen Buddhist Monk* (Kyoto: The Eastern Buddhist Society, 1934; Rutland, Vt.: Charles Tuttle, 1994).

4. Cf. Asataro Miyamori, trans., *An Anthology of Haiku, Ancient and Modern* (Tokyo: Maruzen, 1932), 425.

5. Daisetz T. Suzuki, "The Morning Glory," *The Way* 2, no. 6 (November 1950): 3; and 3, no. 1 (January 1951): 10. Published by The Los Angeles Higashi Hongwanji Young Buddhist Association.

THE LEGACY OF DWIGHT GODDARD

1. Allen Ginsberg, "Negative Capability: Kerouac's Buddhist Ethic," *Tricycle: The Buddhist Review* 2, no. 1 (fall 1992): 8.

2. Barry Gifford and Lawrence Lee, *Jack's Book: An Oral Biography of Jack Kerouac* (New York: St. Martin's Press, 1978), 186.

3. *San Francisco Blues* was composed in 1954 but not published until much later. See Jack Kerouac, *Book of Blues* (New York: Penguin Books USA, 1995), 2–81.

4. Jack Kerouac, *The Dharma Bums* (New York: Viking Press, 1958), 199.

5. David Starry, "Dwight Goddard: The Yankee Buddhist," *Zen Notes* 27, no. 7 (July 1980): 5. Published by the First Zen Institute of America.

6. Daisetz Teitaro Suzuki, *Manual of Zen Buddhism* (Kyoto: The Eastern Buddhist Society, 1935; New York: Grove Press, 1960).

7. For Vasubandhu's seven themes of the *Diamond Sūtra*, see Giuseppe Tucci, *Minor Buddhist Texts*, vol. 1, Series Orientale Roma IX (Rome, Italy: Is.M.E.O., 1956), 24.

8. Dwight Goddard, ed., *A Buddhist Bible* (Boston: Beacon Press, 1994), 447.

9. William F. Powell, trans., *The Record of Tung-shan* (Honolulu: University of Hawaii Press, 1986), 49.

10. A. F. Price, trans., *The Diamond Sutra* and Wong Mou-lam, trans., *The Sutra of Hui-neng* (Boulder, Colo.: Shambhala, 1974).

11. "Inside the FZI, 2: A Buddhist Bible," *Zen Notes* 28, nos. 2–3 (February-March 1981): 4.

12. Starry, "Dwight Goddard," 4.

13. "Inside the FZI, 5," *Zen Notes* 28, no. 7 (July 1981): 2.

14. Ruth Sasaki, trans., *The Recorded Sayings of Lin-chi Hui-chao of Chen Prefecture* (Kyoto: Institute of Zen Studies, 1975); *Zen Dust: The History of the Koan and Koan Studies in Lin-chi (Rinzai) Zen*, trans. with Isshū Miura (New York: Harcourt Brace, 1966).

15. Rick Fields, *How the Swans Came to the Lake* (Boston: Shambhala, 1986), 185.

16. At one point Goddard sought to persuade Christian Science authorities to include some of his translations of Buddhist texts in Mary Baker Eddy's *Science and Health*, but he was refused. See *Zen Notes* 27, no. 7: 5.

17. John Henry Barrows, ed., *The World's Parliament of Religion: The Columbian Exposition of 1893*, vol. 1 (Chicago: Parliament Publishing Company, 1893), 3.

18. Goddard did not speak Japanese and was not knowledgeable enough in Chinese to be able to translate Buddhist texts. Thus he was dependent on native scholars for the primary translation work, notably the

6. Bunnō Kato, Yoshirō Tamura, and Kojirō Miyasaka, with revisions by William Schifter and Pier P. Del Campana, *The Threefold Lotus Sutra: Innumerable Meanings, The Lotus Flower of the Wonderful Law*, and *Meditation on the Bodhisattva Universal Virtue* (New York: Weatherhill, 1975), 319–27.

7. Miura and Sasaki, *Zen Dust*, 203–4.

8. Cf. Thomas Cleary, trans., *Book of Serenity* (Hudson, N.Y.: Lindisfarne Press, 1990), 350.

9. Cf. Thomas Cleary and J. C. Cleary, trans., *The Blue Cliff Record* (Boston: Shambhala, 1992), 554.

10. Robert Aitken, *The Mind of Clover: Essays in Zen Buddhist Ethics* (San Francisco: North Point Press, 1984), 103.

11. Hakuin Zenji, "Song of Zazen," in Robert Aitken, *Encouraging Words: Zen Buddhist Teachings for Western Students* (San Francisco: Pantheon Press, 1993), 180.

12. William Edward Soothill and Lewis Hodous, *A Dictionary of Chinese Buddhist Terms* (London: Kegan Paul, Trench, Trubner, 1937), 328.

13. Cleary and Cleary, *The Blue Cliff Record*, 37.

14. Norman Waddell, trans., "A Selection from the Ts'ai Ken T'an," *The Eastern Buddhist*, New Series 2, no. 2 (1969): 88–89.

15. Cf. *The Gateless Barrier: The Wu-men kuan (Mumonkan)*, trans. with commentaries by Robert Aitken (San Francisco: North Point Press, 1990), 126.

16. Cf. Thomas Cleary, trans., *Transmission of Light* (San Francisco: North Point Press, 1990), 132–34.

17. Cf. Cleary, *Transmission of Light*, 129–31.

18. Cf. Suzuki, *Manual of Zen Buddhism*, 13.

19. "There are only three references to 'mind' in classical Buddhism: (1) *chitta*, thought; (2) *mano*, mind (measuring and comparing); (3) *vinna*, to know." (Conversation with Bhanté H. Gunaratana, Tucson, Ariz., October 3, 1994).

20. John Blofeld, trans., *The Zen Teaching of Huang Po on the Transmission of Mind* (New York: Grove Press, 1958), 36.

21. Yoshito S. Haketa, trans., *The Awakening of Faith: Attributed to Asvaghosha* (New York: Columbia University Press, 1967), 40.

22. See also John Blofeld, *The Zen Teaching of Huang Po*.

23. Philip Kapleau, *The Three Pillars of Zen*, 160.

24. William James, *The Principles of Psychology*, vol. 1 (New York: Henry Holt, 1918), 224–48.

25. George Meredith, "Lucifer in Starlight," in *The Oxford Book of English*

    Verse, 1250–1918, ed. by Arthur Quiller-Couch (Oxford: Clarendon Press, 1939), 960.

26. Cf. Cleary and Cleary, *The Blue Cliff Record*, 274.

27. Sōiku Shigematsu, trans., *A Zen Forest: Sayings of the Masters* (New York: Weatherhill, 1981), 130; cf. 45.

28. Robert Louis Stevenson, *A Child's Garden of Verses* (New York: Charles Scribner's Sons, 1905), 13.

29. Philip B. Yampolsky, trans., *The Platform Sutra of the Sixth Patriarch* (New York: Columbia, 1967), 143.

30. Cf. Aitken, *The Gateless Barrier*, 278.

31. This metaphor appears several times in the writings of Dōgen Kigen. See, for example, Kazuaki Tanahashi, trans., *Moon in a Dewdrop: Writings of Zen Master Dōgen* (San Francisco: North Point Press, 1985), 88. For other references, including those for other metaphors, such as walls, tiles, and stones, see Hee-Jin Kim, *Dōgen Kigen: Mystical Realist* (Tucson: University of Arizona Press, 1987), 111, 273 n. 44.

THE VIRTUE OF ABUSE

1. Dharmas, with a lowercase *d*, can be read "phenomena."

2. Robert Aitken, *Encouraging Words: Zen Buddhist Teachings for Western Students* (San Francisco: Pantheon Books, 1993), 176–77.

3. "Enji" was part of Tōrei's Dharma name, while "Zenji" is his posthumous title.

4. *Tōrei Zenji Bosatsu Gangyō Mon, Zenshū Zaike Nikka Kyō (Sutras Authorized for Daily Use in the Zen Sect)*, Morie Hideji, ed. (Tokyo: Morie Shōten, 1929), n.p.

5. I also acknowledge the help of Anne Aitken and other Diamond Sangha members with the preparation of this essay.

6. These passages are echoes or quotations from Thomas Cleary and J. C. Cleary, trans., *The Blue Cliff Record* (Boston: Shambhala, 1992), 1; Wong Mou-lam, trans., *The Sutra of Hui-neng*, in *The Diamond Sutra and the Sutra of Hui-neng* (Berkeley: Shambhala, 1969), 38; Thomas Cleary, ed. and trans., *The Original Face: An Anthology of Rinzai Zen* (New York: Grove Press, 1978), 102; John Blofeld, trans., *The Zen Teachings of Huang Po on the Transmission of Mind* (New York: Grove Press, 1958), 41; and A. F. Price, trans., *The Diamond Sutra*, in *The Diamond Sutra and the Sutra of Hui-neng*, 45, 29.

7. "To Hear a Sound and Awaken to the Way," in Sonja Arntzen, trans., *Ikkyū and the Crazy Cloud Anthology* (Tokyo: University of Tokyo Press, 1986), 50. References are to Hsiang-yen hearing a stone strike a

bamboo (Robert Aitken, *The Gateless Barrier* [San Francisco: North Point Press, 1990], 39); to Fu of Tai-yüan hearing a bell (Cleary and Cleary, *The Blue Cliff Record*, 328); and to the T'ang period poet T'ao Yüan-ming, who didn't get it (Arntzen, *Ikkyū and the Crazy Cloud Anthology*, 48–51, 181).

8. Butsugen Shōon, in Thomas Cleary, trans. and ed., *Zen Essence: The Science of Freedom* (Boston: Shambhala, 1989), 44.

9. Cf. Cleary and Cleary, *The Blue Cliff Record*, 514.

10. Cleary and Cleary, *The Blue Cliff Record*, 515.

11. Christopher Cleary, trans., *Swampland Flowers: The Letters and Lectures of Zen Master Ta Hui* (New York: Grove Press, 1977), 46–47.

12. Bergan Evans, *Dictionary of Quotations* (New York: Delacorte Press, 1968), 244.

13. Cf. Price, trans., *The Diamond Sutra*, 50. This passage is a kōan, Case Ninety-seven in Cleary and Cleary, *The Blue Cliff Record*, 614.

14. William Edward Soothill and Lewis Hodous, *A Dictionary of Chinese Buddhist Terms* (London: Kegan Paul, Trench, Trubner, 1937), 277.

15. "The Parinirvana Brief Admonitions Sutra," trans. by Kazuaki Tanahashi and Jonathan Condit (unpublished ms., Zen Center of San Francisco, 1980), 5.

16. Yung-chia Hsüan-chüeh, *Cheng-tao ke* (*Shōdōka*, "Song of the Confirmed Tao"), cf. Nyogen Senzaki and Ruth Strout McCandless, *Buddhism and Zen* (San Francisco: North Point Press, 1987), 37–38.

17. W. S. Merwin and Sōiku Shigematsu, trans., *Sun at Midnight: Poems and Sermons by Musō Soseki* (San Francisco: North Point Press, 1989), 37.

18. Aitken, *The Gateless Barrier*, 100.

19. Norman Waddell, trans., *The Unborn: The Life and Teaching of Zen Master Bankei, 1622–93* (San Francisco: North Point Press, 1984), 97–98.

20. Thomas Cleary, trans., *Book of Serenity* (Hudson, N.Y.: Lindisfarne Press, 1990), 163.

21. Penny Lernoux, *Hearts on Fire: The Story of the Maryknoll Sisters* (New York: Orbis Books, 1993), 242–50.

THE WAY OF DŌGEN ZENJI

1. Hee-Jin Kim, *Dōgen Kigen: Mystical Realist* (Tucson: University of Arizona Press, 1987), 100.

2. Simon Pétrement, *Simone Weil: A Life*, trans. by Raymond Rosenthal (New York: Pantheon, 1976), 39–40.

Notes

3. Kim, *Dōgen Kigen*, 100.
4. Kim, *Dōgen Kigen*, 52.
5. This and subsequent passages quoted from the "Genjō Kōan" do not appear in Kim's text and are my own translations, using *Shōbōgenzō*, Honzanban Shukusatsu, ed. (Tokyo: Kōmeisha, 1968).
6. Yamada Kōun Rōshi told this story during a talk at the Koko An Zendō.
7. *The Gateless Barrier: The Wu-men kuan (Mumonkan)*, trans. with commentaries by Robert Aitken (San Francisco: North Point Press, 1990), 54.
8. Dōgen Kigen, *Kyōjukaimon*, cf. Robert Aitken, *The Mind of Clover: Essays in Zen Buddhist Ethics* (San Francisco: North Point Press, 1984), 50.

ULTIMATE REALITY AND THE EXPERIENCE OF NIRVANA

1. Walpola Rahula, *What the Buddha Taught* (New York: Grove Press, 1959), 16–28.
2. *The Gateless Barrier: The Wu-men kuan (Mumonkan)*, trans. with commentaries by Robert Aitken (San Francisco: North Point Press, 1990), 7.
3. See, for example, Kōun Yamada, *Gateless Gate: Newly Translated with Commentary* (Tucson: University of Arizona Press, 1990), 12–16; Zenkei Shibayama, *Zen Comments on the Mumonkan* (New York: Harper and Row, 1974), 19–31.
4. Aitken, *The Gateless Barrier*, 7–9.
5. Cf. Thomas Cleary, trans., *Book of Serenity* (Hudson, N.Y.: Lindisfarne Press, 1990), 335.
6. Isshū Miura and Ruth Fuller Sasaki, *Zen Dust: The History of the Koan and Koan Study in Rinzai (Lin-chi) Zen* (New York: Harcourt Brace & World, 1966), 275.
7. Cf. Thomas Cleary and J. C. Cleary, trans., *The Blue Cliff Record* (Boston: Shambhala, 1992), 449.
8. See Buddhadāsa Bhikkhu, *Me and Mine: Selected Essays*, ed. with an introduction by Donald Swearer (Albany: State University of New York, 1989); Aung San Suu Kyi, *Freedom from Fear: And Other Writings*, ed. by Michael Aris (New York: Viking Penguin, 1991); Maha Ghoshananda, *Step by Step: Meditations on Wisdom and Compassion* (Berkeley: Parallax Press, 1992); A. T. Ariyaratne, *Collected Works*, vol. 1 (Dehiwala, Sri Lanka: Sarvodaya Research Institute, n.d.); Sulak Sivaraksa, *Seeds of Peace: A Buddhist Vision for Renewing Society* (Berkeley: Paral-

lax Press, 1992); Thich Nhat Hanh, *Vietnam: Lotus in a Sea of Fire* (New York: Hill and Wang, 1967); Dalai Lama, *Worlds in Harmony* (Berkeley: Parallax Press, 1992).

9. Tavivat Puntarigvivat, *Bhikkhu Buddhadāsa's Dhammic Socialism in Dialogue with Latin American Liberation Theology* (Ann Arbor: University Microfilms, 1995), 202–3. Originally Ph.D. diss., Temple University, 1994.

RITUAL AND MAKYŌ

1. Robert Aitken, *The Mind of Clover: Essays in Zen Buddhist Ethics* (San Francisco: North Point Press, 1984).

2. Philip Kapleau, ed., *The Three Pillars of Zen: Teaching, Practice, and Enlightenment* (Boston: Beacon Press, 1967), 38–41.

3. Keizan Jōkin, *Zazen Yojinki*, in Thomas Cleary, ed., *Timeless Spring: A Sōtō Zen Anthology* (Tokyo: Weatherhill, 1980), 112–25.

4. Charles Luk, ed., *The Śūrangama Sutra [Leng-yen-ching]* (London: Rider, 1966), 199–236.

5. Robert Aitken, *Taking the Path of Zen* (San Francisco: North Point Press, 1982), 95–96.

6. *The Gateless Barrier: The Wu-men kuan (Mumonkan)*, trans. with commentaries by Robert Aitken (San Francisco: North Point Press, 1990), 160; Thomas Cleary and J. C. Cleary, trans., *The Blue Cliff Record* (Boston: Shambhala, 1992), 218.

KŌANS AND THEIR STUDY

1. Susanne K. Langer, *Philosophy in a New Key: A Study in the Symbolism of Reason, Rite, and Art* (Cambridge, Mass.: Harvard University Press, 1942), 79–102.

2. R. H. Blyth, *Zen in English Literature and Oriental Classics* (Tokyo: Hokuseido Press, 1942), viii.

3. My secretary suggests that Joyce Kilmer is not famous (or infamous) enough to warrant mention. His poem "Trees," however, remains the best-known bad poem in the English language.

4. *The Gateless Barrier: The Wu-men kuan (Mumonkan)*, trans. with commentaries by Robert Aitken (San Francisco: North Point Press, 1990), 28.

5. Blyth, *Zen in English Literature*, 75. In his translation, Blyth renders *uma* as "cob," an unfamiliar English word that best describes the short, stocky Japanese horse.

Notes

6. Asataro Miyamori, *Haiku: Ancient and Modern* (Tokyo: Maruzen, 1932), facing p. 155.

7. Robert Pinsky, "The Muse in the Machine; Or, The Poetics of Zork," *New York Times Book Review*, 19 March 1995, p. 26.

8. Thomas Cleary, *Entry into the Inconceivable: An Introduction to Hua-yen Buddhism* (Honolulu: University of Hawaii Press, 1983), 9.

9. Cf. Thomas Cleary and J. C. Cleary, trans., *The Blue Cliff Record* (Boston: Shambhala, 1992), 341.

10. I have heard that a certain Western teacher claims that when he understood one kōan, he understood five hundred of them. That's not the way the process works. Understanding—"standing under"—one kōan, you can see the way to resolve five hundred, but explicitly they are not clear and must be personalized one by one.

11. Philip Kapleau, ed., *The Three Pillars of Zen: Practice, Teaching, and Enlightenment* (Boston: Beacon Press, 1965), 204–8. The quotation is originally from Dōgen Kigen, "Busshō," *Shōbōgenzō*. See Hee-Jin Kim, *Dōgen Kigen: Mystical Realist* (Tucson: University of Arizona Press, 1987), 122.

12. Cleary and Cleary, *The Blue Cliff Record*, 342. See Aitken, *The Gateless Barrier*, 46.

13. Robert Aitken, *Encouraging Words: Zen Buddhist Teachings for Western Students* (San Francisco: Pantheon Press, 1993), 173.

14. The Ten (sometimes Six) Ox-herding Pictures depict steps on the Zen Buddhist path to full realization. See D. T. Suzuki, *Manual of Zen Buddhism* (New York: Grove Press, 1960), 120–44.

15. Isshū Miura and Ruth Fuller Sasaki, *Zen Dust: The History of the Koan and Koan Study in Rinzai (Lin-chi) Zen* (New York: Harcourt Brace & World, 1966), 46–76.

16. Cf. Cleary and Cleary, *The Blue Cliff Record*, 412.

17. Cleary, *Entry into the Inconceivable*, 33.

18. Aitken, *The Gateless Barrier*, 283.

19. Thomas Cleary, trans., *Book of Serenity* (Hudson, N.Y.: Lindisfarne Press, 1990), 173.

20. Aitken, *The Gateless Barrier*, 283.

21. Cleary and Cleary, *The Blue Cliff Record*, xvi–xvii.

22. Heinrich Zimmer, *The King and the Corpse: Tales of the Soul's Conquest of Evil*, ed. by Joseph Campbell (Princeton University Press, 1973).

23. Cleary and Cleary, *The Blue Cliff Record*, 37.

24. Aitken, *The Gateless Barrier*, 3.

25. Aitken, *Encouraging Words*, 173.

26. Cleary and Cleary, *The Blue Cliff Record*, 110.
27. A line from Wu-men Hui-k'ai, cited in Aitken, *The Gateless Barrier*, 248.

MARRIAGE AS SANGHA

1. E. E. Cummings, "this little bride & groom are," *Complete Poems 1904–1962* (New York: Liveright Publishing, 1991), 470.
2. Wendell Berry, "Poetry and Marriage," *Standing by Words* (San Francisco: North Point Press, 1983), 200.
3. I gave this talk at the Koko An Zendō in 1986 and would now expand it to include gay and lesbian couples.

DEATH: A ZEN BUDDHIST PERSPECTIVE

1. Isshū Miura and Ruth Fuller Sasaki, *Zen Dust: The History of the Koan and Koan Study in Rinzai (Lin-chi) Zen* (New York: Harcourt Brace & World, 1966), 206.
2. Miura and Sasaki, *Zen Dust*, 326.
3. Miura and Sasaki, *Zen Dust*, 170–71.
4. Miura and Sasaki, *Zen Dust*, 305.
5. Miura and Sasaki, *Zen Dust*, 171.
6. Philip Kapleau, ed., *The Three Pillars of Zen: Teaching, Practice, and Enlightenment* (Boston: Beacon Press, 1967), 173.
7. *The Gateless Barrier: The Wu-men kuan (Mumonkan)*, trans. with commentaries by Robert Aitken (San Francisco: North Point Press, 1990), 219.
8. Aitken, *The Gateless Barrier*, 279.
9. Thomas Cleary, trans., *Shōbōgenzō: Zen Essays by Dōgen* (Honolulu: University of Hawaii Press, 1986), 118.
10. Thomas Cleary, trans., *Book of Serenity* (Hudson, N.Y.: Lindisfarne Press, 1990), 56.
11. Kapleau, *The Three Pillars of Zen*, 168–69.
12. Philip Larkin, *Collected Poems*, ed. by Anthony Thwaite (New York: Farrar Straus Giroux and Marvell Press, 1988), 208–9.
13. Simone Weil, *Gateway to God*, ed. by David Raper, with the collaboration of Malcolm Muggeridge and Vernon Sproxton (New York: Crossroad, 1974), 48.
14. Robert Aitken, *Encouraging Words: Zen Buddhist Teachings for Western Students* (San Francisco: Pantheon Press, 1993), 174.
15. Hakuin Ekaku, "Song of Zazen," in Aitken, *Encouraging Words*, 179.

16. Hee-Jin Kim, *Dōgen Kigen: Mystical Realist* (Tucson: University of Arizona Press, 1987), 17.

17. Cf. Yoel Hoffman, ed., *Japanese Death Poems: Written by Zen Monks and Haiku Poets on the Verge of Death* (Rutland, Vt.: Tuttle, 1986), 145.

18. Cf. Lewis McKenzie, trans., *The Autumn Wind: A Selection from the Writings of Issa* (Tokyo: Kodansha, 1984), 5.

19. Yün-men Wen-yen, cf. Thomas Cleary and J. C. Cleary, trans., *The Blue Cliff Record* (Boston: Shambhala, 1992), 37.

20. R. H. Blyth, *A History of Haiku* (Tokyo: Hokuseido Press, 1963–64), retranslated from the Japanese text, 363.

THE PATH BEYOND NO-SELF

1. A. F. Price, trans., *The Diamond Sutra*, in *The Diamond Sutra and the Sutra of Hui Neng* (Berkeley: Shambhala, 1969), 29, 35, 41–42, 73.

2. For a discussion of Mauthner's thought, see Charles B. Maurer, *Call to Revolution: The Mystical Anarchism of Gustav Landauer* (Detroit: Wayne State University Press, 1972), 58–66.

3. Thomas Cleary and J. C. Cleary, trans., *The Blue Cliff Record* (Boston: Shambhala, 1992), 536.

4. Maurer, *Call to Revolution*, 98.

5. Cleary and Cleary, *The Blue Cliff Record*, 554.

6. Chapter Thirty-nine of the *Mahāparinirvāna Sūtra*, not, so far as I know, translated into English. See Dōgen, *Bendōwa*, "On the Endeavor of the Way," in Kazuaki Tanahashi, *Moon in a Dewdrop: Writings of Zen Master Dōgen* (San Francisco: North Point Press, 1985), 153–54 and Glossary under "Senika," 328.

7. Maurer, *Call to Revolution*, 71.

8. Gustav Landauer, *For Socialism*, trans. by David J. Parent (St. Louis: Telos Press, 1978), 40.

9. *Plato: Gorgias*, trans. Donald J. Zeyl (Indianapolis: Hacket Publ. Co., 1987).

10. Martin Buber, *Paths in Utopia* (New York: Macmillan, 1988), 48.

11. Maurer, *Call to Revolution*, 89.

12. Landauer, *For Socialism*, 39.

13. Landauer, *For Socialism*, 39–40.

14. Landauer, *For Socialism*, 25.

ENVISIONING THE FUTURE

1. E. F. Schumacher, *Small Is Beautiful: Economics As If People Mattered* (New York: Harper & Row, 1975), 55.

2. A. T. Ariyaratne, *Collected Works*, vol. 1 (Dehiwala, Sri Lanka: Sarvodia Research Institute, n.d.); Sulak Sivaraksa, *A Buddhist Vision for Renewing Society: Collected Articles by a Concerned Thai Intellectual* (Bangkok: Thai Watana Panich, 1981).

3. I originally used "Dhamma," the Pali orthography, rather than "Dharma," out of deference to my Theravada listeners.

4. Wes Jackson, *Altars of Unhewn Stone: Science and the Earth* (San Francisco: North Point Press, 1987), 126.

5. Schumacher, *Small Is Beautiful*, 55. A woman's work blesses us and equally our products as well! Schumacher wrote his words before male writers finally learned that the term "man" is not inclusive.

6. Donald K. Swearer, "Three Legacies of Bhikkhu Buddhadāsa," in *The Quest for a New Society*, ed. by Sulak Sivaraksa (Thai Interreligious Commission for Development; Santi Pracha Dhamma Institute, 1994), 17. Cited from Buddhadāsa Bhikkhu, *Buddhasāsanik Kap Kān Anurak Thamachāt* [Buddhists and the Conservation of Nature] (Bangkok: Komol Keemthong Foundation, 1990), 34.

7. James Hillman, "And Huge Is Ugly." Tenth Annual E. F. Schumacher Memorial Lecture, Bristol, England, November 1988.

8. Charles B. Maurer, *Call to Revolution: The Mystical Anarchism of Gustav Landauer* (Detroit: Wayne State University Press, 1972), 58–66. For Spanish origins and developments of the Grupo de Afinidad, see *The Anarchist Collectives: Workers' Self-Management in the Spanish Revolution 1936–1939*, ed. by Sam Dolgof (New York: Free Life Editions, 1974).

9. Mev Puleo, *The Struggle Is One: Voices and Visions of Liberation* (Albany: State University of New York, 1994), 14, 22, 25, 29.

10. Thomas Cleary, *Entry into the Inconceivable: An Introduction to Huayen Buddhism* (Honolulu: University of Hawaii Press, 1983), 7.

11. William Foote Whyte and Kathleen King Whyte, *Making Mondragon: The Growth and Dynamics of the Worker Cooperative Complex* (Ithaca, N.Y.: ILR Press, Cornell University, 1988), 3, 30. Other cooperatives worthy of study include the Transnational Information Exchange, which brings together trade unionists in the same industry across the world; the Innovation Centers, designed in Germany to help workers who must deal with new technologies; and Emilia Romagna in northern Italy, networks of independent industries that research and market products jointly. Jeremy Brecher, "Affairs of State," *The Nation* 260, no. 9, 6 March 1995, p. 321.

12. After presenting this paper, I learned about Tavivat Puntarigvivat's

Ph.D. dissertation at Temple University, 1994: *Bhikkhu Buddhadāsa's Dhammic Socialism in Dialogue with Latin American Liberation Theology* (Ann Arbor, University Microfilms, 1995).

13. Carl J. Bellas, *Industrial Democracy and the Worker-Owned Firm: A Study of Twenty-one Plywood Companies in the Pacific Northwest* (New York: Praeger Publishers, 1972).

14. Peter Stiehler, "The Greed of Usury Oppresses," *The Catholic Agitator* 24, no. 7 (November 1994): 5.

15. Jill Torrie, ed., *Banking on Poverty: The Global Impact of the IMF and World Bank* (Toronto: Between the Lines, 1983), n.p.

16. Torrie, *Banking on Poverty*, 14. See also Doug Bandow and Ian Vásquez, eds., *Perpetuating Poverty: The World Bank, the IMF, and the Developing World* (Washington, D.C.: Cato Institute, 1994), and Kevin Danaher, *Fifty Years Is Enough: The Case Against the World Bank and the IMF* (Boston: South End Press, 1994).

17. The Bretton Woods system of international currency regulation was established at the United Nations Monetary and Financial Conference, representing forty-five countries, held at Bretton Woods, New Hampshire, in July 1944. The United States dollar was fixed to the price of gold and became the standard of value for all currencies.

18. Noam Chomsky, *The Prosperous Few and the Restless Many* (Berkeley: Odonian Press, 1993), 6.

19. Gore Vidal, "The Union of the State," *The Nation* 259, no. 22, 26 December 1994, p. 789.

20. I use "Siam" rather than "Thailand" to honor the position taken by progressive Buddhists in that country, who point out that the Thais are only one of their many ethnic peoples and that the new name was imposed by a Thai autocrat.

21. Abu N. M. Wahid, *The Grameen Bank: Poverty Relief in Bangladesh* (Boulder, Colo.: Westview Press, 1993).

22. See, for example, Ivan Light and Edna Bonacich, *Immigrant Entrepreneurs: Koreans in Los Angeles, 1965–1982* (Berkeley, Los Angeles, London: University of California Press, 1988), 244.

23. Paul Glover, "Creating Economic Democracy with Locally Owned Currency," *Terrain*, December 1994, pp. 10–11. See also "An Alternative to Cash: Beyond Banks or Barter," *New York Times*, 31 May 1993, p. 8, and "The Potential of Local Currency," by Susan Meeker Lowrey, *Z Magazine*, July-August 1995, pp. 16–23.

24. Nejatullah Siddiqui, *Banking Without Interest* (Delhi: Markazi Maktaba Islami, 1979), x–xii.

A

25. Robert Aitken, *Encouraging Words: Zen Buddhist Teachings for Western Students* (San Francisco: Pantheon Press, 1993), 179.

26. Michael Phillips and Sallie Rashberry, *Honest Business: A Superior Strategy for Starting and Conducting Your Own Business* (New York: Random House, 1981).

27. Real Goods, for example, retailers of merchandise that helps to sustain the habitat. Address: 966 Mazzoni Street, Ukiah, CA 95482-0214. Catalog for March 1995, p. 37.

28. One does feel this urgency in the literature of Real Goods. Let us hope this remarkable company is a forerunner of others.

29. Sulak, *A Buddhist Vision for Renewing Society*, 108.

THE EXPERIENCE OF EMPTINESS

1. A. F. Price, trans., *The Diamond Sutra*, in *The Diamond Sutra and The Sutra of Hui-neng* (Berkeley: Shambhala, 1969), 29, 32, 41–42, 65. For a further discussion of the *Diamond Sūtra*, see the essay "The Path Beyond No-Self" in this volume.

2. John Blofeld, trans., *The Zen Teachings of Huang-po* (New York: Grove Press, 1958), 41.

3. Cf. Thomas Cleary, *Book of Serenity* (Hudson, N.Y.: Lindisfarne Press, 1990), 241.

4. D. T. Suzuki, *Zen and Japanese Culture* (Princeton, N.J.: Princeton University Press, 1970), 73. The Kamakura era (1192–1333) was a time of cultural reformation in Japan.

5. Cf. Kazuaki Tanahashi, trans., *Moon in a Dewdrop: Writings of Zen Master Dōgen* (San Francisco: North Point Press, 1985), 70.

6. Suzuki, *Zen and Japanese Culture*, 115. See Robert Aitken, *The Mind of Clover: Essays in Zen Buddhist Ethics* (San Francisco: North Point Press, 1984), 5–6.

7. Kenneth Morgan, ed., *The Path of the Buddha: Buddhism Interpreted by Buddhists* (New York: Ronald Press, 1956), 309–10.

8. *Bhagavad Gita*, vol. 2, 17–19, cited in Aitken, *The Mind of Clover*, 179.

9. Mike Sayama, *Samurai: Self-Development in Zen, Swordsmanship, and Psychotherapy* (New York: State University of New York, 1986), 65–74. On page 67, Sayama quotes his teacher Tenshin Tanouye: "In Zen, after you go through your kōan training, you ask yourself what is your … frame of mind going through life. Miyamoto Musashi said, 'It is like a huge boulder rolling downhill.' I say, if you like to ride your Honda bike through life, that's OK, but I'm riding my tank."

Notes

10. Robert Aitken, *A Zen Wave: Bashō's Haiku and Zen* (New York: Weatherhill, 1978), 74–79.

11. Cf. Haku'un Yasutani, "Introductory Lectures on Zen Training," in Philip Kapleau, ed., *The Three Pillars of Zen: Teaching, Practice, and Enlightenment* (Boston: Beacon Press, 1969), 28.

12. See "The Brahma Vihāras" in this volume. Also, Robert Aitken, *The Practice of Perfection: The Pāramitās from a Zen Buddhist Perspective* (San Francisco: Pantheon Books, 1994), 93, and Har Dayal, *The Bodhisattva Doctrine in Sanskrit Literature* (Delhi: Motilal Banarsidas: 1934), 225–29.

13. See "The Virtue of Abuse" in this volume.

14. Cf. R. H. Blyth, *Zen in English Literature and Oriental Classics* (New York: Dutton, 1960), 158.

BRAHMADANDA

1. I shared earlier drafts of this paper with many people and consulted on the subject with several others. Despite all this help, I take full responsibility for the points made.

2. Bergen Evans, *Dictionary of Quotations* (New York: Delacorte Press, 1968), 286–87. Bradford himself was later led to the stake, a consummation that added poignancy to his piety.

3. *The Gateless Barrier: The Wu-men kuan (Mumonkan)*, trans. with commentaries by Robert Aitken (San Francisco: North Point Press, 1990), 133.

4. *The Survivor Activist*, Frank and Sara Fitzpatrick, 52 Lyndon Road, Cranston, R.I. 02905-1121; telephone: (401) 491-2548.

5. Bhanté H. Gunaratana, "The Nature of Reality," a talk given at a symposium of the same title, Arizona Teachings, Tucson, Arizona, October 2, 1994.

6. Maurice Walshe, trans., *Thus Have I Heard: The Long Discourses of the Buddha* (London: Wisdom Publications, 1987), 270. For the use of brahmadanda as heavenly retribution, see *Ambattha Sutta* 1.23; Walshe, *Thus Have I Heard*, 117, 156 n, 549.

7. Robert Aitken, *The Mind of Clover: Essays in Zen Buddhist Ethics* (San Francisco: North Point Press, 1984), 43.

8. Vernon E. Johnson, *Intervention: How to Help Someone Who Doesn't Want Help—A Step-by-Step Guide for Families and Friends of Chemically Dependent Persons* (Minneapolis: Johnson Institute, 1986).

9. William Stafford, "Thinking for Berky," in *The Darkness Around Us Is*

Deep: Selected Poems by William Stafford, ed. by Robert Bly (New York: HarperCollins, 1993), 37.

10. The second case, the story of an old woman who burned down the hut of a monk who responded inappropriately to sexual advances, may be found in Aitken, *The Mind of Clover*, 38–39.

11. Ryōmin Akizuki, *New Mahāyāna: Buddhism for a Postmodern World* (Berkeley: Asian Humanities Press, 1990), 45.

ABOUT MONEY

1. Thomas Cleary and J. C. Cleary, trans., *The Blue Cliff Record* (Boston: Shambhala, 1992), 187.

2. *The Gateless Barrier: The Wu-men kuan (Mumonkan)*, trans. with commentaries by Robert Aitken (San Francisco: North Point Press, 1990), 273; Thomas Cleary, *Book of Serenity* (Hudson, N.Y.: Lindisfarne Press, 1990), 335.

HERALD BIRDS

1. Yaichiro Isobe, trans., *Musings of a Chinese Vegetarian* (Tokyo: Yuhodo, 1926), 199. Translated from the Chinese and Japanese, which are included in the book. See also William Scott Wilson, trans., *Roots of Wisdom: Saikontan, by Hung Ying-ming* (New York: Kodansha International, 1985), 101.

2. Cited in Jonathan Cott, *Wandering Ghost: The Odyssey of Lafcadio Hearn* (New York: Kodansha International, 1992), 268–69, from Lafcadio Hearn, *Glimpses of Unfamiliar Japan* (Boston & New York: Houghton Mifflin Co., 1894).

3. Cf. Thomas Cleary and J. C. Cleary, trans., *The Blue Cliff Record* (Boston: Shambhala, 1992), 309.

4. Cf. Katsuki Sekida, trans., *Two Zen Classics: Mumonkan and Hekiganroku* (New York: Weatherhill, 1977), 147. Unfortunately Mr. Sekida treats Pai-chang's weeping and laughing simply as emotional relief.

WALLACE STEVENS AND ZEN

1. References to Stevens's poetry are accompanied by the page number in *The Collected Poems of Wallace Stevens* (New York: Knopf, 1954). The abbreviation "L." followed by a page number refers to *Letters of Wallace Stevens*, ed. by Holly Stevens (New York: Knopf, 1972). Unattributed translations are the author's.

2. Cf. Thomas Cleary and J. C. Cleary, trans., *The Blue Cliff Record* (Boston: Shambhala, 1992), 306.

3. Cf. Kazuaki Tanahashi, trans., *Moon in a Dewdrop: Writings of Zen Master Dōgen* (San Francisco: North Point Press, 1985), 69.

4. Cf. Cleary and Cleary, *The Blue Cliff Record*, 554.

5. Cf. Tanahashi, *Moon in a Dewdrop*, 69.

6. Cf. Cleary and Cleary, *The Blue Cliff Record*, 176.

7. Robert Aitken, *Encouraging Words: Zen Buddhist Teachings for Western Students* (San Francisco: Pantheon Press, 1993), 171.

8. Cf. Thomas Cleary, trans., *Transmission of Light* (San Francisco: North Point Press, 1990), 219.

9. Cf. Cleary and Cleary, *The Blue Cliff Record*, 1.

10. Cf. *The Gateless Barrier: The Wu-men kuan (Mumonkan)*, trans. with commentaries by Robert Aitken (San Francisco: North Point Press, 1990), 284.

11. D. T. Suzuki, *Essays in Zen Buddhism*, second series (London: Rider, 1950), 248.

12. Cf. Zenkei Shibayama, *Zen Comments on the Mumonkan* (New York: Harper & Row, 1974), 10.

13. Robert Aitken, *A Zen Wave: Bashō's Haiku and Zen* (New York and Tokyo: Weatherhill, 1978), 84.

PLAY

1. Cf. Thomas Cleary, trans., *Book of Serenity* (Hudson, N.Y.: Lindisfarne Press, 1990), 367.

2. Cf. Ruth Fuller Sasaki, trans., *The Recorded Sayings of Ch'an Master Lin-chi Hui-chao of Chen Prefecture* (Kyoto: Institute for Zen Studies, 1975), 50–52.

3. *The Gateless Barrier: The Wu-men kuan (Mumonkan)*, trans. with commentaries by Robert Aitken (San Francisco: North Point Press, 1990), 7–18. "Mu" is the Japanese pronunciation of the pertinent ideograph, and "Wu" is the contemporary Mandarin pronunciation. However, I am told that "Mu" was probably the pronunciation of Chao-chou's time.

4. Cf. Cleary, *Book of Serenity*, 76.

5. Sōiku Shigematsu, trans., *A Zen Forest* (New York: Weatherhill, 1981), 100.

6. Cf. D. T. Suzuki, *Essays in Zen Buddhism: Third Series* (New York: Samuel Weiser, 1976), 86.

7. Isshū Miura and Ruth Fuller Sasaki, *Zen Dust: The History of the Koan*

and Koan Study in Rinzai (Lin-chi) Zen (New York: Harcourt Brace and World, 1966), 112.

8. Francis Beaumont, *The Knight of the Burning Pestle*, ed. by Andrew Gunn (Berkeley: University of California Press, 1968).

9. Robert Aitken, ed., "Miscellaneous Kōans" (Diamond Sangha, Honolulu, mimeographed), n.p.

1. Hee-Jin Kim, *Dōgen Kigen: Mystical Realist* (Tucson: University of Arizona Press, 1987), 61–62.

2. D. T. Suzuki, *Manual of Zen Buddhism* (New York: Grove Press, 1960), 134 and plate 11.

3. Suzuki, *Manual of Zen Buddhism*, 129.

4. *The Gateless Barrier: The Wu-men kuan (Mumonkan)*, trans. with commentaries by Robert Aitken, (San Francisco: North Point Press, 1990), 248.

5. Cf. Thomas Cleary and J. C. Cleary, trans., *The Blue Cliff Record* (Boston: Shambhala, 1992), 386. It was said at that time that until one had interviewed the National Teacher, one's practice was not complete.

6. The circle is the perennial base of temple architecture. Anthony Lawlor, *The Temple in the House: Finding the Sacred in Everyday Architecture* (New York: G. T. Putnam's Sons, 1994), 116–20.

7. *Sauvastika* is distinguished from *svastika* (swastika) in Sanskrit. See William Edward Soothill and Lewis Hodous, *A Dictionary of Chinese Buddhist Terms* (London: Kegan Paul, Trench, Trubner, 1937), 203. However, sauvastika has not found its way into English usage. *The Oxford English Dictionary* includes the meaning of "sauvastika" under "swastika."

8. Cf. Thomas Cleary, trans., *Book of Serenity* (Hudson, N.Y.: Lindisfarne Press, 1990), 324. Wan-sung, editor of the *Book of Serenity*, explains that Rucika, incarnated as the guardians at the gate of monasteries, was a disciple of the Buddha who "vowed to attain the ornaments of skill of 999 Buddhas." He is thus the archetype of protecting the Dharma. Cleary, *Book of Serenity*, 327.

9. See Glossary. Also: William F. Powell, trans., *The Record of Tung-shan* (Honolulu: University of Hawaii Press, 1986), 61–63; "The *Goi* Kōans [with a commentary by Hakuin Ekaku]," in Isshū Miura and Ruth Fuller Sasaki, *Zen Dust: The History of the Koan and Koan Study in Rinzai (Lin-chi) Zen* (New York: Harcourt Brace & World, 1966), 67–71.

10. Cleary and Cleary, *The Blue Cliff Record*, 603. Other examples of circles and ideographs in Zen literature: the nun Shih-chi (Jissai) circling the seat of Chü-chih (Gutei), Aitken, *The Gateless Barrier*, 29; Ma-ku circling the seats of Chang-ching (Shōkei) and Nan-ch'üan, Cleary and Cleary, *The Blue Cliff Record*, 194; and twice Tzu-fu (Shifuku) drawing circles in the air, once with an ideograph within it, Cleary and Cleary, *The Blue Cliff Record*, 206, 581.

A Glossary of Buddhist Names, Terms, and Usages

Most Japanese and Chinese names are given in the traditional order, with surnames first. The terms that are italicized in definitions are also entries. For Mahayana, read Mahayana Buddhist or Buddhism. For Zen, read Zen Buddhist or Buddhism. For Rinzai or Sōtō, read Rinzai or Sōtō Zen Buddhist or Buddhism. Note that Sanskrit terms that conventionally begin with Ś will be found under *Sh*. Abbreviations: c=century; C=Chinese; J=Japanese; P=Pali; S=Sanskrit.

affinity. The tendency of *beings* to come together as organisms, families, species, and other groupings, providing individuality and diversity within the *plenum*.

ahimsā. (S). "Nonharming," the first *precept*.

Amida (J). See *Amitābha*.

Amitābha (S); *Amida* (J). The *Buddha* of Infinite Light and Life. *Archetype* of transformation and salvation in *Pure Land* schools.

Ānanda, 4th c., B.C.E. One of the principal disciples of the *Buddha Shākyamuni*; the second *ancestral teacher*.

ancestral teachers. Teachers in the traditional *Zen* lineage. *Founding teachers, patriarchs.*

anger. An emotional response to something that is inappropriate or unjust. An emotion involved in self-protection. See *hatred*.

anguish. In Buddhism, painful resistance to the reality of mortality and dependence. See *duhkha*.

225

Glossary

antinomianism. In Buddhism: the notion that one can ignore the *precepts.* Reckless freedom.

anuttara-samyak-sambodhi (S); *daigo tettei* (J). Greatly enlightened to the very bottom. Full and complete *realization.* Thorough accomplishment of the third of the *Four Noble Truths.*

archetype. In Buddhism: a metaphor empowered by innate *understanding* and long-term usage. A legendary or historical figure who models an empowered metaphor and who can be made one's own.

arhat (S); *arahant* (P). One who has destroyed the obstacles to *nirvana;* the *Classical Buddhist* ideal.

Avalokiteshvara (S). Sovereign observer. Archetypal *Bodhisattva* of mercy. See *Kanzeon.*

Awakening of Faith; Ta-ch'eng ch'i-hsin lun (C). Seminal *Mahayana* exposition of *Buddha Nature.*[1]

Bankei Yōtaku (J), 1622–93. Zen master of the *Rinzai* school.

barrier. In Zen, a checkpoint, as at a frontier.

Bashō, Matsuo (J), 1644–94. First great haiku poet; associated with *Zen.*

Bassui Tokushō (J), 1327–78. Zen master of the *Rinzai* school.

beings. All entities that exist. *Sentient beings.* See *many beings.*

bhikkhu (P). Mendicant monk in *Classical Buddhism.*

Blue Cliff Record; Pi-yen lu (C); **Hekiganroku** (J). A 13th-c. collection of one hundred Zen cases with comments; associated with the *Rinzai* school.[2]

bodhi (S). *Enlightenment.*

Bodhi tree. Ficus Religiosa. The tree that sheltered the Buddha *Shākyamuni* before, during, and just after his *realization.*

bodhichitta (S). The aspiration for *enlightenment* and *Buddhahood.*

Bodhidharma (S), 6th c. Semilegendary Indian or West Asian founder of *Ch'an Buddhism; archetype* for steadfast *practice.*

bodhimanda (S). The spot or place under the *Bodhi tree* where the Buddha *Shākyamuni* had his realization. *Dōjō.*

Bodhisattva (S). One on the *path* to *enlightenment;* one who is enlightened; one who enlightens others; a figure in the Buddhist pantheon.

Bodhisattva precepts. See *precepts.*

Bodhisattva vows. Basic vows of the *Mahayana* Buddhist: (1) to save all beings, (2) to abandon obstacles, (3) to waken to or *understand* the many dharmas, (4) to attain the *Buddha Way.*

---

[1] Yoshito S. Haketa, trans., *The Awakening of Faith Attributed to Asvaghosha* (New York: Columbia University Press, 1967).

[2] Thomas Cleary and J. C. Cleary, trans., *The Blue Cliff Record* (Boston: Shambhala, 1992).

Glossary

"Bodhisattva's Vow"; "Bosatsu Gangyō Mon" (J). A homily by Tōrei Enji. Distinguish from *"Bodhisattva vows."*

body and mind dropped away. The *self* forgotten in *zazen* or other activity.

Book of Serenity; Ts'ung-jung lu (C); **Shōyōroku** (J). A 13th-c. Chinese collection of one hundred Zen cases with comments; associated with the *Sōtō* school.[3]

Brahma Vihāra (S). Sublime Abode. The four progressive Brahma Vihāras are *maitrī*, boundless loving-kindness; *karunā*, boundless compassion; *mudita*, boundless joy in the liberation of others; and *upekshā*, boundless equanimity.

Buddha (S). Enlightened One. *Shākyamuni*. An enlightened person. A figure in the Buddhist pantheon. Any *being*.

Buddha Dharma (S). The teaching of the Buddha *Shākyamuni* and his successors; *Dharma*; Buddhism; the *Eightfold Path*; *Buddha Tao or Way*.

Buddhadāsa Bhikkhu, 1906–93. Thai Buddhist master, founder of a progressive movement of Thai Buddhism.

Buddha Nature. Essential nature, self-nature, or *true nature*.

Buddha Tao or Way. Buddha Dharma. The *Eightfold Path*.

Buddhahood. Enlightenment and compassion. The condition of a *Buddha*.

cause and effect. One explanation of *karma*.

Ch'an (C). See *Zen*.

Ch'ang-sha Ch'ing ts'en (C); *Chōsa Keijin* (J), d. 868. Ch'an master in the *Nan-yüeh line*.

Chao-chou Ts'ung-sheng (C); *Jōshū Jūshin* (J), 778–897. Ch'an master in the *Nan-yüeh line*.

Cheng-tao ke (C); **Shōdōka** (J). "Song of Realizing the Way." A long Dharma poem by Yung-chia.[4]

Chih-i or *T'ien-t'ai Chih-i*, 538–97. Founder of the T'ien-t'ai (C) Tendai (J) school of Mahayana Buddhism.

Ch'ing-yüan Hsing-ssu (C); *Seigen Gyōshi* (J), d. 740. Ch'an master, successor of *Hui-neng*, founder of the line that became the Ts'ao-tung (*Sōtō*), *Fa-yen*, and *Yün-men* schools of Ch'an and (in the case of Sōtō) *Zen*.

Ch'ing-yüan line. See *Ch'ing yüan Hsing-ssu*.

Chiyo-ni or *Chiyo-jo* (J). 1701–75. Poet and nun.

Chü-ti (C); *Gutei* (J), late-9th-c. Ch'an master in the *Nan-yüeh line*.

Classical Buddhism. The Buddhism that preceded the rise of the *Mahayana*. Modern *Theravada*.

[3] Thomas Cleary, trans., *Book of Serenity* (Hudson, N.Y.: Lindisfarne Press, 1990).
[4] Sheng-yen, *The Sword of Wisdom: Lectures on "The Song of Enlightenment"* (Elmhurst, N.Y.: Dharma Drum Publications, 1990).

Glossary

clinging. A preoccupation with the *self* and with the notion of permanence. The source of *duhkha*. The second of the *Four Noble Truths*.

confirmation. In Zen: affirmation of *realization* by one's teacher. Experientially, *realization* is itself confirmation.

cross over. *Pāramitā*; *save*; transform.

Daiō Kokushi; Nampo Jōmyō (J), 1235–1309. Zen master, a founder of the *Rinzai* school in Japan.

dai oshō (J). Great priest. A posthumous honorific.

daishi (J). Great master. A posthumous title.

dāna (S). Charity, giving, relinquishment (and their perfections). The *First Pāramitā*.

darani (J). See *Dhāranī*.

dedication; ekō (J). Turning. Transferring one's *merit* to another. Transferring the merit of a *sūtra* recitation to *Buddhas, Bodhisattvas, ancestral teachers*, and so on.

denominator (coinage by Yamada Kōun). *Essential nature* underlying and infusing all things. See *numerator*.

dhāranī (S); *darani* (J). An invocation of praise.

Dharma (S); *Dhamma* (P). Religious, secular, or natural law; the *law of karma*; *Buddha Dharma or Tao*; teaching; the *Dharmakāya*. With a lower-case *d*: a phenomenon or thing.

Dharma gates. Incidents or *particulars* that can enable one's *realization*. The various teachings.

Dharmakāya (S). The *Dharma* or law body of pure and clear, *essential nature*. See *Three Bodies of the Buddha*.

Dharma transmission. Formal empowerment by the old master of a new teacher in the traditionally unbroken line of masters from the Buddha *Shākyamuni*.

dhyāna (S). Focused meditation and its form. *Zazen, Zen*. See *samādhi*.

Diamond Sangha. A network of Zen centers founded in Honolulu in the *Sanbō Kyōdan* tradition.

Diamond Sūtra; Vajrachedikā Sūtra (S). A text of the *Mahāprajñāpāramitā literature* that stresses freedom from concepts.

discursive (English-language usage). Explanatory, prosaic, not presentational.

Dō (J). See *Tao*.

Dōgen Kigen (J), 1200–53. Zen master, founder of the *Sōtō* school in Japan, author of the *Shōbōgenzō*.

dōjō (J); *bodhimanda* (S). The training hall or *zendō*. One's own place of *realization*.

dokusan; sanzen (J). To work alone; personal interview with the *rōshi* during formal *practice*.

dukkha (P); *duhkha* (S). *Anguish*; a response to mortality and dependence. The consequences of denying that reality. The first of the *Four Noble Truths*.

ego. In Buddhism: self-image. *Self.* Distinguish from selfish and egocentric.

Eightfold Path. The ideals and *practice* of Right Views, Right Thoughts, Right Speech, Right Conduct, Right Livelihood, Right Effort or Life-style, Right Recollection, and Right Absorption or Concentration—in keeping with the insubstantial nature of the *self, mutual interdependence*, and the sacred nature of each *being*. The way of freeing oneself from *duhkha*. The fourth of the *Four Noble Truths*.

emptiness, empty. The insubstantial nature of the *self* and all selves. Realized as the same as substance.

engaged Buddhism (coinage by Thich Nhat Hanh). Taking the *path*, especially as a community. *Practice* within or alongside poisonous systems.

enlightenment. Bodhi. The ideal condition of *realization*.

essential nature. The pure and clear *void* that is charged with potential. The *denominator* of phenomena or *beings. Self-nature, true nature, Buddha Nature.*

evil. Harmful, destructive. Distinguish from immoral as dogma.

Fa-yen (C) school; *Hōgen* (J) school. Distinguished by the tendency of teachers to throw students' questions back to them. Founded by Fa-yen Wen-i (C)—Hōgen Bun'eki (J)—885–958, Ch'an master in the *Ch'ing-yüan line*.

Five Modes of the Particular and the Universal. A poetical work by *Tung-shan Liang-chieh* that recaps the insights of Zen *practice*.[5]

Five Precepts. See *Pañcha Shīla.*

Five Skandhas. See *skandha.*

forgetting the self. Body and mind dropped (or fallen) *away.* The experience of everything disappearing with an act or with something sensed. Might be confirmed as *realization*.

founding teachers. See *ancestral teachers.*

Four Abodes. See *Brahma Vihāras.*

Four Noble Truths. The basic Buddhist teaching: *anguish* is everywhere; *clinging* is the cause of anguish; there is *liberation* from anguish; the *Eightfold Path* is the *way* of this liberation.

---

[5]William F. Powell, *The Record of Tung-shan* (Honolulu: University of Hawaii Press, 1986), 61–63.

gasshō (J); *añjali* (S). The *mudra* of hands held palm to palm before the lower part of the face, in devotion, gratitude, or as a greeting.

gateless. Completely open.

Gateless Barrier, The, Wu-men kuan (C); **Mumonkan** (J). A 13th-c. collection of forty-eight Zen cases, compiled with commentaries by *Wu-men.*[6]

gāthā (S). A four-line verse that sums up an aspect of the *Dharma.* In the *Mahayana* it is often a *vow.*

Gautama (S); *Gotama* (P). Personal name of the Buddha *Shākyamuni.*

"Genjō Kōan." "The Fundamental Kōan Actualized." A chapter of the *Shōbōgenzō.*

genkan (J). Entry to a Japanese home or temple.

Goddess of Mercy. See *Kanzeon.*

greed. Affinity exploited to serve the self. First of the *Three Poisons.*

Gyōgi Bosatsu (J), 670–749. A Buddhist master and civil engineer, possibly of Korean origin.

Hakuin Ekaku (J), 1685–1768. Zen master of the *Rinzai* school and an artist.

harmony. Interdependent co-arising realized.

hatred. Indulging or dwelling in *anger.* Second of the *Three Poisons.*

Heart Sūtra; Prajñā Pāramitā Hrdaya Sūtra (S); **Hannya Shingyō** (J). A brief summary of the *Mahāprajñāpāramitā literature,* stressing the complementarity of substance and emptiness.[7]

Hotei (J); *Putai* (C). The "Laughing Buddha," associated with *Maitreya.* Archetype of fulfilled *realization.*

Hsiang-lin Teng-yüan (C); *Kōrin Chōen* (J), d. 987. Ch'an master in the *Yün-men* school.

Hsüeh-t'ou Ch'ung-hsien (C); *Setchō Jūken* (J), 982–1052. Ch'an master of the *Yün-men* school, compiler of the *Blue Cliff Record.*

Huang-po Hsi-yün (C); *Ōbaku Ki'un* (J), d. 850. Ch'an master in the *Nanyüeh* line.

Hua-yen (C); *Kegon* (J). Teachings found in the *Hua-yen ching* and its commentaries.

Hua-yen ching or **Hua-yen Sūtra.** Chinese version of the *Avatamsaka Sūtra,* which stresses the *particularity* of all *beings* and their innate *harmony.*[8]

[6] *The Gateless Barrier: The Wu-men kuan (Mumonkan),* trans. with commentaries by Robert Aitken (San Francisco: North Point Press, 1990).

[7] Robert Aitken, *Encouraging Words: Zen Buddhist Teachings for Western Students* (San Francisco: Pantheon Books, 1993), 173–75.

[8] Thomas Cleary, trans., *The Flower Ornament Scripture: A Translation of the Avatamsaka Sutra,* 3 vols. (Boulder: Shambhala, 1984–87).

Glossary

Hui. Ch'an monk, unknown except as a colleague of *Ch'ang-sha.*

Hui-k'o (C); *Eka* (J), 487–593. Early Ch'an master, traditionally the successor of *Bodhidharma.*

Hui-neng (C); *Enō* (J), 638–713. Sixth *ancestral teacher.* Traditionally the key figure in Ch'an Buddhist acculturation. Founder of *Southern Ch'an.*

Hung-chih Cheng-chüeh (C); *Wanshi Sogaku* (J), 1091–1157. Compiler of the *Book of Serenity.*

ignorance. Neglecting or ignoring *essential nature,* the primal *harmony* of *beings,* and their sacredness. Third of the *Three Poisons.* Distinguish from *not knowing.*

Ikkyū Sōjun (J), 1394–1481. Zen master of the *Rinzai* school and a poet.

Inka Shōmei (J). Legitimate seal of clearly furnished proof. The affirmation and document(s) of *Dharma* transmission.

interbeing (coinage by Thich Nhat Hanh). The *Sambhogakāya.* The many as the self or as the particular, the dynamics of that reality, and its experience.

interdependent co-arising. The function and dynamics of *interbeing.* Mutual *interdependence.*

intimacy. In Zen: the nature of *practice* and its experience.

I-hsin chieh-wen (C); *Isshin Kaimon* (J), *Precepts of one mind.* Attributed to Bodhidharma, but probably by *Chih-i.*

Issa Kobayashi (J), 1763–1827. Haiku poet, associated with *Pure Land* views and practices.

kalpa (S). A particular aeon. An immeasurably long period of time.

Kan-feng (C); *Kempō* (J), 9th c. Ch'an master in the *Nan-yüeh line.*

Kannon (J). See *Kanzeon.*

Kanzan Kokushi; Kanzan Egen (J), 1277–1360. Zen master of the *Rinzai* school.

Kanzeon, Kannon (J); *Kuan-yin* (C). One Who Perceives the Sounds of the World; the archetypal *Bodhisattva* of mercy. Derived from *Avalokiteshvara.*

karma (S). Action. *Cause and effect; affinity;* the function of *interdependent co-arising.* Interdependence. Distinguish from fixed fate.

karunā (S). Boundless compassion. See *Brahma Vihāra.*

Kegon (J). See *Hua-yen.*

kenshō (J). Seeing *(true) nature. Realization.* See *satori.*

kinhin (J). Walking verification. *Sūtra* walk. The formal walk between periods of *zazen.*

knowledge. Formulated *wisdom.*

kōan (J). *Universal/particular.* A *presentation* of the harmony of the universal and the particular; a theme of *zazen* to be made clear. A traditional Zen story.

Kuan-yin (C). See *Kanzeon*.

Kuei-shan Ling-yü (C); *Issan Reiyū* (J), 771–853. Ch'an master in the *Nan-yüeh line*, cofounder with *Yang-shan* of the Kuei-yang (Igyō) school of Ch'an, noted for harmony between teacher and disciple.

Kuei-yang school. See *Kuei-shan*.

Kuei-tsung Chih-ch'ang (C); *Kisu Chijō* (J), 8th–9th c. Ch'an master in the *Nan-yüeh line*.

Kumārajiva, 344–413. Central Asian Buddhist master instrumental in translating important Buddhist texts into Chinese.

Lankavatara Sūtra. *Mahayana* text that expounds the *Middle Way*.[9]

law of karma. The way things act. *Karma. Cause and effect, affinity, interdependent co-arising*.

Layman P'ang; P'ang-yün (C); *Hō Koji* (J), 740–808. Lay Ch'an master in the *Ch'ing-yüan* and *Nan-yüeh lines*.

liberation. Freedom from *clinging*. The third of the *Four Noble Truths*. *Realization. Prajñā*.

life and death or *birth and death*. *Samsāra*; the realm of transience, relativity, and *karma*.

Lin-chi I-hsüan (C); *Rinzai Gigen* (J), d. 866. Ch'an master, founder of the Lin-chi (*Rinzai*) school.

Ling-yün Chih-ch'in (C); *Reiun Shigon* (J), 9th c. Ch'an master in the *Nan-yüeh line*.

Lotus Land. Nirvana. Pure Land.

Lotus Sūtra; Saddharma Pundarīka Sūtra (S). A devotional and metaphysical text presented in allegorical form.[10]

Mahākāshyapa; Mahākāsyapa, 4th c., B.C.E. Principal heir of the Buddha *Shākyamuni*. The first *ancestral teacher*.

Mahāparinirvāna Sūtra (S); **Mahāparinibbāna Sutta** (P). Differing accounts of the Buddha *Shākyamuni*'s last days, his last teachings, and his death.[11]

[9]Daisetz Teitaro Suzuki, trans., *The Lankavatara Sūtra* (London: George Routledge and Sons, 1932).

[10]Bunnō Katō, Yoshirō Tamura, and Kojirō Miyasaka, with revisions by William Schiffer and Pier Del Campana, trans., *The Threefold Lotus Sutra* (New York: Weatherhill, 1975).

[11]Kosho Yamamoto, trans., *Mahaparinirvana-sutra: A Complete Translation from the Classical Chinese, in 3 Volumes* (Ube-shi, Yamaguchi-ken: Karinbunko, 1973–75). Distinguish from the shorter Theravada text: *Mahāparinibbāna Sutta*, in Maurice Walshe, trans., *Thus I Have Heard: The Long Discourses of the Buddha* (London: Wisdom Publications, 1987), 231–90.

Mahāprajñāpāramitā (S) *literature*. Six-hundred-fascicle exposition of the *Middle Way*. See *Heart Sūtra, Diamond Sūtra*.

mahāsattva (S). Great noble being.

Mahayana, Mahayāna (S). Great Vehicle; the Buddhism that arose five hundred years after *Shākyamuni*. The Buddhist tradition of China, Japan, and Korea; also found, together with *Theravada*, in Vietnam. Tibetan Buddhism is often considered to be Mahayana. The *practice* of saving the *many beings*.

Maitreya (S). The Compassionate One; the future, potential, or inherent *Buddha*.

maitrī (S); *mettā* (P). See *Brahma Vihāra*.

Ma-ku. See *Pao-che of Ma-ku*.

makyō (J). Uncanny realm. A deep dream of participation in the *Buddha Dharma*. Distinguish from sensory, visual, or auditory distortion.

Mañjushrī; Mañjuśrī (S). Beautiful Virtue; archetypal *Bodhisattva* of *wisdom*.

mantra (S). An empowered phrase or text.

many beings; shujō (J). All *beings*. Distinguish from *sentient beings*.

Māra (S). The destroyer; the evil one.

Ma-tsu Tao-i (C); *Baso Dōitsu* (J), 709–88. Ch'an master, successor of *Nan-yüeh*.

Maudgalyāyana, 4th c., B.C.E. Prominent disciple of the Buddha *Shākyamuni*.

merit. The good results of good action. A function of *karma* and *interdependent co-arising*.

metaphor. A *presentation* of something in terms of another, expressing their unity. In Buddhism: any presentation.

metta (P); *maitrī* (S). Boundless loving-kindness. See *Brahma Vihāra*.

Middle Way. The *Way* of the Buddha. Harmonizing the *particular* and the universal, *cause and effect, essential nature* and *phenomena*, the *Three Bodies of the Buddha*, and so on. The *Eightfold Path*. Moderation.

Mind. The unknown and unknowable that comes forth as the *plenum* with its particular *beings* and their interdependence and affinities. *Essential nature*. The human mind.

mindfulness, mindful. Attention. Attention to the breath or task. *Right Recollection* of the *self* and others as insubstantial, interdependent, and sacred. See *Eightfold Path*.

Miyamoto Musashi (J), 1584–1645. Samurai, artist, and professional hero.

mondō (J). Question and answer, the Zen dialogue. See *dokusan, kōan*.

morality, moral. Refers to the process of character formation and the state of personal nobility. Pursued on the *Eightfold Path* and fulfilled in the *pāramitās*.

Mu (J); *Wu* (C). No; does not have. A *kōan* from Case One of *The Gateless Barrier*.

muditā (S). Boundless joy in the liberation of others. See *Brahma Vihāra*.

mudra (S). A seal or sign; hand or finger position, gesture, or posture that presents an aspect of the *Dharma*.

Musō Soseki (J), 1275–1351. Zen master of the *Rinzai* school and a poet.

mutual interdependence. Interdependent co-arising.

Nakagawa Sōen (J), 1907–84. Zen master of the *Rinzai* school and a poet.

Nan-ch'üan P'u-yüan (C); *Nansen Fugan* (J), 749–835. Ch'an master in the *Nan-yüeh* line.

Nantō (J). Difficult case. A category of kōans.

Nan-yüeh Huai-jang (C); *Nangaku Ejō* (J), 677–744. Ch'an master, successor of *Hui-neng*, founder of the line that became the *Kuei-yang* and *Lin-chi* (*Rinzai*) schools of Ch'an and (in the case of Rinzai) Zen.

Nan-yüeh line. See *Nan-yüeh Huai-jang.*

National Teacher Chung; Nan-yang Hui-chung (C); *Chū Kokushi; Nanyō Echū* (J), d. 776. Ch'an master, successor of *Hui-neng*.

Net of Indra. A model of the *plenum* found in *Hua-yen* teaching. Each point in the net contains all the other points.

Nichiren sect (J). A *Mahayana* tradition based on the *Lotus Sūtra*, founded by Nichiren, 1222–82.

Nirmānakāya (S). The transformation body of uniqueness and variety. See *Three Bodies of the Buddha*.

nirvana; nirvāna (S). Extinction of craving; *liberation* found in *practice* and *realization*. See *Pure Land*.

noble, nobility. In Buddhism: keeping the Buddha *Dharma*. Faultless demeanor.

Northern school. A Ch'an tradition descending from *Shen-hsiu*. The so-called gradual school, as distinguished from the *Southern* or "sudden school."

no-self. A peak experience of *shūnyatā* disclosing the vanity, futility, and ignorance of self-centeredness.

not knowing. Accepting the fact of mystery at the essence.

numerator (coinage by Yamada Kōun). The phenomenal aspect of reality. See *denominator*.

oshō (S). Priest, a senior monk (title).

Ox-herding Pictures. A traditional rendering of the progressive steps on the Zen Buddhist path.

Pai-chang Huai-hai (C); *Hyakujō Ekai* (J), 720–814. Ch'an master in the *Nan-yüeh line*, traditionally the founder of the Ch'an/Zen monastic tradition.

Pañcha Shīla. The first *Five Precepts*, universally accepted by both lay and ordained Buddhists.

Pao-che of Ma-ku; Maku (or *Mayü*) *Pao-che* (C); *Mayoku Hōtetsu* (J), 8th c. Ch'an master in the *Nan-yüeh line.*

pāramitā (S). Perfection as condition or *practice. Cross over* (to the shore of *nirvana*). *Save;* transform. The *Six Pāramitās* are the ideals of charity, morality, patience, vitality, absorption or concentration, and wisdom. In another enumeration, the Ten Pāramitās include, in addition, skillful means, aspiration, strength of purpose, and knowledge.

Parinirvāna Brief Admonitions Sūtra. One of the *Mahāpārinirvana Sūtras,* devoted to the doings and teachings of the Buddha toward the end of his life.[12]

particularity or *particular.* In Zen: the nature of a *phenomenon* or a *being.*

Path. The *Eightfold Path. Tao, Way, Buddha Dharma.*

patriarchs. Ancestral teachers.

pattica-samuppāda (P). *Interdependent co-arising.* Mutual interdependence. The function of *interbeing.*

perfection. See *pāramitā.*

phenomena, phenomenon. Beings, a *being.* The *numerator* of *essential nature.*

Platform Sūtra of the Sixth Ancestor; Liu-tsu t'an-ching (C); **Rokuso Dankyō** (J). Memorializes *Hui-neng.*[13]

plenum. The *universe* and its *many beings.* Realized as the *void.*

practice; shugyo (J). Austerities, training. Endeavors in the *dōjō; zazen.* To take the *Eightfold Path;* to follow the *precepts;* to *turn the Dharma wheel.*

prajñā (Prajñā Pāramitā) (S). *Wisdom, enlightenment,* or *bodhi* (and their perfection).

precepts. In the Mahayana, the Sixteen Bodhisattva Precepts are the Three Vows of refuge in the *Three Treasures;* The Three Pure Precepts of avoiding *evil,* practicing good, and saving the *many beings;* and the Ten Grave Precepts of not killing, not stealing, not misusing sex, not lying, not giving or taking drink or drugs, not speaking of faults of others, not praising oneself while abusing others, not sparing the *Dharma* assets, not indulging in *anger,* and not defaming the *Three Treasures.*

presentation, presentational (English-language usage). A *particular* expression or appearance without *discursive* explanation.

---

[12] "The Parinirvana Brief Admonitions Sutra," trans. by Kazuaki Tanahashi and Jonathan Condit (unpublished ms., Zen Center of San Francisco, 1980).

[13] Philip B. Yampolsky, *The Platform Sutra of the Sixth Patriarch: The Text of the Tunhuang Manuscript* (New York: Columbia University, 1967).

Glossary

Pure Land. Nirvana; the afterlife envisioned in the Pure Land schools of Buddhism. Lotus Land. Realized as this very place.

Pure Land Buddhism. Faith in the saving power of the Buddha *Amitābha*. Practice of personal transformation.

Pu-tai (C). See *Hōtei*.

realization; genjō (J). Actualization, personalization. A glimpse of empty or unitive possibilities of wholeness. *Prajñā* experienced through one of the senses, acknowledged by a confirmed teacher. Made true for one-self. *Kenshō*. Understanding. Confirmation. The third of the *Four Noble Truths*.

rebirth. The coherent but changing *karma* of an individual or a cluster of individuals reappearing after death. The continuous arising of coherent, changing karma during life. Distinguish from *reincarnation*, or rebirth of the body.

reincarnation. The notion that an enduring self reappears after death in a new birth. Distinguish from *rebirth*.

Right Recollection. Mindfulness of ephemerality, *interbeing*, and the sacred nature of all things. See *mindfulness*, *Eightfold Path*.

rinpoché. Tibetan master (a title).

Rinzai Zen Buddhism. Today the Zen school in which *kōan* study is used in conjunction with *zazen*.

rōshi (J). Old teacher. Now the title of the confirmed Zen teacher.

sage. In Buddhism: an enlightened, compassionate person. A *Buddha*.

samādhi (S). Absorption. The quality of *zazen*. One with the *universe*. See *dhyāna*.

Samantabhadra (S). Pervading Goodness. Archetypal *bodhisattva* of great action (in *turning the Dharma wheel*).

Samatha (P); *Shamatha* (S). Concentration. The Chinese translate it by its mental function: "stopping." The meditative practice of stillness that reveals the insubstantial nature of the self.

Sambhogakāya (S). The bliss body of *mutual interdependence*. See *Three Bodies of the Buddha*.

samsāra (S). The rising and falling of *life and death*. The relative world, realized as the same as *nirvana*.

samu (J). Work ceremony. Temple maintenance as part of formal *practice*.

Sanbō Kyōdan. Order of the *Three Treasures*. A lay Japanese *Sōtō* school that includes elements of *Rinzai* practice, founded by *Yasutani Haku'un* in Kamakura, Japan.

sangha (S). Aggregate. A community or all communities of ordained Buddhists. Lay Buddhist community or communities. Any community, including that of all *beings*.

satori (J). *Prajñā*, *enlightenment*; the condition or experience of enlightenment. See *realization*, *kenshō*.

save. In Buddhism: enable or help (someone) to *cross over* to full *realization*. Transform (someone or something) for the better.

self. In Buddhism: the insubstantial individual that is nonetheless unique and sacred.

self-nature. The essential quality of the *self*. *True nature*, *essential nature*, *Buddha Nature*.

Seng-ts'an (C); *Sōsan* (J), d. 606? Ch'an master, traditionally the successor of *Hui-k'o.* Said to be the author of the *Hsin-hsin ming* (C; Shinjinmei [J]), a long *Dharma* poem.

sensei (J). Teacher.

sentient beings; ujō (J). *Beings* with senses. Human beings.

Senzaki Nyogen, 1876–1958. *Rinzai* monk and teacher.

sesshin (J). To touch, receive, and convey the *Mind*. The intensive Zen retreat of three to seven days.

Shaku Sōen (J), 1859–1919. Zen master of the *Rinzai* school.

Shākyamuni (S). Sage of the Shākya clan; the historical *Buddha*, 5th–4th c. B.C.E. Founder of Buddhism. *Archetype* of *prajñā*, *karunā*, and the *nirmānakāya*.

Shāriputra (S), 4th c., B.C.E. A prominent disciple of the Buddha *Shākyamuni*, interlocutor in the *Heart Sūtra*.

Shen-hsiu (C); *Jinshū* (J), 605?–706. Successor of *Hui-neng*, brother of Hui-neng in the *Dharma* and founder of *Northern* Ch'an.

shikan (J). *Shamantha/Vipashyanā*. The meditation of the *T'ien-t'ai* school. See *T'ien-t'ai, shikantaza, zazen*.

shikantaza (J). *Body and mind dropped away* in *zazen*.

shīla, śīla (S). Restraint. Morality; keeping the *precepts*.

Shintō (J). Way of the Gods. The indigenous religion of Japan, distinguished by a veneration of nature.

shinzanshiki (J). "The ceremony of ushering into the mountain"; ceremony of becoming an abbot.

Shōbōgenzō (J). **True Dharma Eye Treasure**. The collection of *Dōgen*'s many essays on Zen and its practice.

Shōtoku Taishi (J), 573–621. Prince regent of Japan, instrumental in introducing Buddhism to Japan.

shūnyatā, śūnyata (S). The *void* that is charged with potential.

Shūrangama Sūtra. Best known of sūtras with this title is Chinese in origin. It sets forth the many delusions and the means for seeing through them.[14]

sit, sitting. Zazen.

Six Pāramitās. See *pāramitā.*

Sixteen Bodhisattva Precepts. See *precepts.*

Sixth Ancestor or *Patriarch. Hui-neng.*

skandha (S). Aggregate. The five skandhas that make up the *self* are forms of the world, sensation, perception, mental reaction, and consciousness. Realized as *empty.*

"Song of Zazen"; "Zazen Wasan" (J). *Dharma* poem by Hakuin.[15]

Sōtō Zen Buddhism. Today the Zen sect that uses *shikantaza* as a principal practice in *zazen.*

Southern school. The enduring school of Ch'an, descending from *Hui-neng.* The so-called sudden school.

Sudhāna (S). Protagonist in Book Three of the *Hua-yen ching.*

suffering. Enduring, allowing; enduring pain. Distinguish from *anguish.* See *duhkha.*

Süng-yüan Ch'ung-yüeh (C); *Shōgen Sugaku* (J), 1139–1209. Ch'an master of the *Lin-chi* school.

Śūrangama Sūtra. See **Shūrangama Sūtra.**

sutra; sūtra (S); *sutta* (P). Sermons by the Buddha *Shākyamuni* and those attributed to him; Buddhist scripture. See *Tripitaka.*

Suzuki, D. T., 1870–1966. The *Rinzai* lay scholar most responsible for the early dissemination of knowledge about Zen in the Americas and Europe.

Suzuki Shunryū, 1904–71. *Sōtō* master, founder of the Zen Center of San Francisco.

swaraj (Hindi). Self-government, political and personal.

taking refuge. The ceremony of acknowledging the *Buddha, Dharma,* and *Sangha* as one's home, common to all Buddhist traditions. See *precepts.*

Takuan Sōhō (J), 1573–1645. Zen master of the *Rinzai* school.

Takuhatsu (J); *Pindapāta* (S). Ceremonial acceptance of alms in the neighborhood of the temple.

Ta-lung Chih-hung (C); *Dairyū Chikō* (J), 9th c. Ch'an master in the *Ch'ing yüan* line.

tan (J). Row; line of people doing *zazen. Dōjō.*

---

[14]Charles Luk, trans., *The Śūrangama Sūtra (Leng Yen Ching)* (London: Rider, 1966).

[15]Robert Aitken, *Encouraging Words: Zen Buddhist Teachings for Western Students* (San Francisco: Pantheon Books, 1993), 179–80.

Glossary

Tan-yüan Ying-chen (C); *Tangen Ōshin* (J), 8th–9th c. Ch'an master, successor of *National Teacher Chung*.

Tao (C). *Way*. In Buddhism: the Buddha *Dharma*; the *Eightfold Path*.

Tao-hsin (C); *Dōshin* (J), 580–651. Ch'an master, traditionally the successor of *Seng-ts'an*.

Tao-te ching; Tao-teh-king. Central text of Taoism, attributed to Lao-tzu.

T'ao Yüan-ming (C); *Tōenmei* (J), 365–427. Poet.

Ta-sui Fa-chen (C); *Daizui Hōshin* (J), 8th–9th c. Ch'an master in the *Nan-yüeh line*.

Tathāgata (S). One who comes forth (presenting *essential nature* with particular qualities). A *Buddha. Shākyamuni*.

Ta-yü; Kao-an Ta-yü (C); *Kōan Daigu* (J), 8th–9th c. Ch'an master in the *Nan-yüeh line*.

teishō (J). Presentation of the shout; the *Dharma* presented by the *rōshi* in a public talk.

Ten Grave Precepts. See *precepts*.

Tendai (J). See *T'ien-t'ai*.

Theravada; Theravāda (P). Way of the elders. Today the Buddhism of South and Southeast Asia. See *Classical Buddhism*.

Three Bodies (of the Buddha). The complementary natures of Buddhahood and the world: *Dharmakāya*, the *Dharma* or law body of *essential nature*; *Sambhogakāya*, the bliss body of *mutual interdependence*; and *Nirmānakāya*, the transformation body of uniqueness and variety.

Three Poisons. Greed, hatred, and ignorance: the main kinds of self-centeredness that hinder the *practice*. See *klesha*.

Three Pure Precepts. See *precepts*.

Three Treasures or *Jewels*. The *Buddha, Dharma*, and *Sangha; enlightenment*, the *Way*, and community—the basic elements of Buddhism.

Three Vows of Refuge. See *taking refuge, precepts*.

Three Worlds. Realms of consciousness: desire, form, and no-form—that is, attachment, acceptance, and transcendence. Also past, present, and future.

T'ien-t'ai (C); *Tendai* (J). A school of Buddhism that includes scholastic, devotional, esoteric, and meditative teachings. An antecedent of Ch'an/Zen and other schools.

Ting-chou Shih-tsang (C); *Jōshu Sekizo* (J), 714–800. *Northern school* Ch'an master.

Tōrei Enji (J), 1721–92. Zen master of the *Rinzai* school.

T'ou-tzu I-ching (C); *Tōsu Gisei* (J), 1021–83. Ch'an master of the Ts'ao-tung (*Sōtō*) school.

Glossary

Tripitaka (S). Three baskets. The three main teachings of Buddhism: *sūtras*, the *vinaya*, and the abhidharma (commentaries).

true nature. Self, essential, or *Buddha Nature*.

Ts'ai-ken tan (C); *Saikontan* (J). A Ming-period book of Confucian, Buddhist, and Taoist homilies by Hung Ying-ming (C; Kōjisei [J]).

Ts'ao-ch'i (C). *Hui-neng's* temple. *Hui-neng*.

Ts'ung-jung lu (C). *Book of Serenity*.

Tung-shan Liang-chieh (C); *Tozan Ryokai* (J), 807–69. Ch'an master, founder of Ts'ao-tung (*Sōtō*) school of Ch'an/Zen.

Tung-shan Shou-ch'u (C); *Tōzan Shushu* (J), 910–90. Ch'an master in the *Yün-men* school.

turning the Dharma wheel. Lending *wisdom* and energy to the transformational process of the *Buddha Dharma* in the world. See *practice, engaged Buddhism*.

Tusita Heaven (S). The fourth of the six heavens in the realm of desire where *Maitreya* Buddha lives.

understanding. Stepping under, and taking on oneself. Knowing.

universe. The *plenum*. The *void*.

upāya (S). Skillful, appropriate means of *turning the Dharma wheel* or prompting *realization*.

upekshā (S). Boundless equanimity. See *Brahma Vihāra*.

Vairochana, Vairocana (S). The Sun Buddha. *Archetype* of *bodhi*, total purity, and the *Dharmakāya*.

Vajrayana; Vajrayāna (S). The Way of the Adamantine Truth; Tibetan Buddhism.

Vasubandhu, 4th or 5th c. Indian Buddhist master and philosopher.

vinaya (S). The moral teachings. See *Tripitaka*.

Vipassanā (P); *Vipashyanā, Vipaśyanā* (S). Insight. The meditative practice of seeing into the insubstantial nature of the self and its sensations, thoughts, and emotions.

void. Shūnyatā. Vast *emptiness* that is full of potential. See *mind, Dharmakāya*. Realized as the *plenum*.

Vow. Usually the expression of resolve to attain Buddhahood and that all beings attain it. See *bodhichitta*.

Wan-sung Hsing-hsin or *Lao-jen* (C); *Banshō Gyōshu* or *Rōjin* (J), 1166–1246. Ch'an master of the Ts'ao-tung school, editor of the *Book of Serenity*.

Way. Tao, Dharma.

wheel of the Dharma. The evolution of the *Buddha Dharma* in universal consciousness. See *turning the Dharma wheel*.

Wisdom. Prajñā. Realization and its insights.

Wu-chiu (C); *Uchū* (J), 8th–9th c. Ch'an master, a successor of *Ma-tsu* in the *Nan-yüeh* line.

Wu-cho (C); *Mugaku* (J), 821–900. Ch'an master of the *Kuei-yang* (Igyō) *school* of Ch'an.

Wu-men Hui-K'ai (C); *Mumon Ekai* (J), 1183–1260. Chinese master in the *Lin-chi* (*Rinzai*) school, compiler of *The Gateless Barrier*.

Yamada Kōun, 1907–89. Japanese master of the *Sanbō Kyōdan*.

Yamamoto Gempō (J), 1866–1961. Zen master of the *Rinzai* school.

Yamaoka Tesshū (J), 1836–88. Swordsman and statesman.

Yang-shan Hui-chi (C); *Kyōzan Ekaku* (J), 807–83. Ch'an master in the *Nan-yüeh* line, cofounder with *Kuei-shan* of the *Kuei-yang* (Igyō) *school* of Ch'an.

Yasutani Haku'un, 1895–1973. Zen master, founder of the *Sanbō Kyōdan*.

Yüan-wu K'o-ch'in (C); *Engo Kokugon* (J), 1063–1135. Ch'an master of the *Lin-chi* school, editor of the *Blue Cliff Record*.

Yüeh-shan Wei-yen (C); *Yakusan Igen* (J), 8th and 9th c. Ch'an master in the *Ch'ing-yüan* line.

Yung-chia Hsüan-chüeh (C); *Yoka Genkaku* (J), 665–713. Ch'an master, successor of *Hui-neng*, author of *Cheng-tao ke*.

Yün-men Wen-yen (C); *Unmon Bun'en* (J), 864–949. Ch'an master in the *Ch'ing-yüan* line, founder of the Yün-men school of Ch'an, one of the precursors of modern *Rinzai*.

zazen (J). The practice of seated, focused meditation.

zazenkai (J). Zen group. A brief *sesshin*.

Zen (J); *Ch'an* (C). Focused, exacting meditation; the Zen tradition.

zendō (J). Zen hall; Zen center. *Dōjō*.

zenji (J). Zen master, usually a posthumous honorific. Also, monk.

Printed in the United States
by Baker & Taylor Publisher Services